BIRDS
and their World

BIRDS
and their World

John Andrews

Hamlyn
London·New York·Sydney·Toronto

Published by
The Hamlyn Publishing Group Limited
London · New York · Sydney · Toronto
Astronaut House, Feltham, Middlesex, England
Copyright ©The Hamlyn Publishing Group Limited 1976

ISBN 0 600 34883 0

Filmset in England by Siviter Smith & Co. Ltd., Birmingham
Printed in Holland

Contents

Introduction

The first problem that faces all creatures is how to stay alive. All living things at all times are under the pressures put on them by the habitats in which they live and by the other creatures with which they compete. These pressures change more or less slowly and individuals or species which cannot adapt to them die.

Birds have been in the past, and still remain, a very successful group of creatures. There are over 8 000 species living in the world today. They occupy an enormous variety of habitats. Some drink nectar in the constant summer of the tropical forests, others incubate their eggs in the continuous darkness and horrifying cold of an Antarctic mid-winter. Some soar without apparent effort in the thin air high above the Andean peaks, others pursue their prey for long minutes deep beneath the cold waters of the

North Sea. All are products of at least 150 million years of evolution, during which countless other species have developed, flourished and become extinct.

The fossil history of birds is relatively little known, mainly because they are usually small and have a rather delicate bone structure which rarely produces good fossils. However, a few early remains do provide clues from which it is possible to make guesses at the course of avian evolution.

Birds evolved from reptiles. Medium-sized, arboreal lizards, perhaps not unlike species alive today, found it advantageous to jump from tree to tree rather than descend to the ground where they would be especially vulnerable and exposed to predators. Gradually a limited gliding ability could have been evolved which would have allowed safe crossing of wider gaps between trees

Left
As soon as they are hatched, the female red-breasted merganser leads the chicks to water where they can feed themselves and are less vulnerable to predators than if all were in the nest together. Mergansers feed on fish, their beaks having a bluntly serrated edge to help grasp them.

Above right
The hoatzin is a particularly unusual bird of the South American rainforest. Living by the rivers, it feeds largely on aquatic vegetation. It is a capable swimmer and has claws on its wings to aid it in clambering about in the waterside shrubs.

and had a clear survival value. Individuals fortuitously born with elongated scales—the precursors of feathers—on the forelimb would have had a marginal advantage over those without, as this would create a better shape for gliding. Thus there evolved over millions of years creatures which, though reptilian in body shape and features, possessed the birdlike ability to glide even if they could not yet make powered flights.

Here guesswork links up with reality for in 1861 the fossil remains of *Archaeopteryx* were found in Bavaria in fine Jurassic sedimentary rock. A second, more complete skeleton was found in 1877 and two more in 1956. *Archaeopteryx* was a pigeon-sized bird with short round wings and a long tail from which feathers stuck out laterally unlike those of modern birds. It had a reptilian, toothed head and its skeleton was too flexible to sustain powerful flight.

We must come forward another fifty million years before more fossil remains are available, particularly from the shale beds of Kansas. These include *Ichthyornis*, a gull-sized bird with well-developed wings and a keeled breast bone; this bone is the point of attachment for flight muscles and its size indicates that the bird was a capable flier. From the same epoch comes *Hesperornis*, a large cormorant-like species apparently specialized for swimming and diving with much-reduced wings.

Another forty million years takes us to the Eocene, when it is possible to detect the beginnings of types that exist today including

Above
Displaying before the female, a male mute swan partly spreads his wings over his back and arches his neck. Though swans mate for life, courtship is important as it strengthens the pair bond and helps to start the physiological processes on which breeding depends.

herons, ducks, hawks, waders and owls. In the intervening sixty million years that bring us to the present, birds have undergone an adaptive radiation, continuing to adjust and refine their specialized adaptations in order to cope with changing climates and vegetations, and latterly with the problems and opportunities made by man.

The thing that is unique about birds is their covering of feathers, which not only gives them the ability to fly extremely well but also carries out a number of other important functions. Feathers provide a covering that retains warmth, repels water and helps to protect the body from physical damage. The outer coat of body feathers are called contour feathers and they form a continuous covering enclosing an inner layer of down; air trapped in the down between contour feathers and skin insulates the body and minimizes heat loss. Fluffing up the feathers increases insulation but in hot weather birds may have to pant to keep cool.

The most important result of this is that birds are able to be warm-blooded creatures, unlike reptiles which are cold-blooded. This means that they function efficiently at air temperatures in which reptiles would become torpid and that they can therefore exploit habitats in colder climates

Left
A red-throated diver sits on its nest close to the water's edge. With their long necks, long beaks and streamlined bodies, divers are efficient underwater hunters of fish, coming ashore only to nest. Most summer on inland waters in tundra and semi-tundra habitats, wintering at sea in coastal areas.

Above
The American turkey vulture feeds largely on carrion and is one of the few birds believed to have a keen sense of smell. Most vultures rely on sharp eyesight to detect their food but the turkey vulture also has very large nostrils and olfactory organs to aid its search.

Left
A great reed warbler feeds
a young cuckoo. By laying
eggs in other species' nests,
the cuckoo can probably get
more young reared than it
could feed by its own efforts
and the adults can leave
before their insect food
supply becomes short in the
autumn.

Above
The chiffchaff is an
insectivorous warbler
feeding in deciduous and
coniferous woodland in
Europe and Asia and largely
moving south to Africa and
India in winter, though a few
overwinter in milder parts.
The nest is usually built off
the ground in thick
vegetation and the insect
food is gathered from trees
and bushes.

and seasons than reptiles can tolerate. Obviously, to fulfil this and their other functions, feathers must be in good condition and thus require regular preening and periodic replacement by moulting.

Feathers have also developed other specialized uses. Nightjars and some other birds have hair-like feathers around the base of the beak to facilitate the capture of insect prey in flight. Birds such as herons and bitterns, whose plumage may become slimy in dealing with prey such as eels, have powder-down–feathers which disintegrate to yield a fine powdery substance used to dry up soiled areas and aid preening.

Feathers may be coloured and sometimes

Right
A number of birds of prey feed on fish and the osprey is probably the best known. Usually hunting over lakes it drops to the surface to grasp in its talons any large fish swimming near the surface, often without wetting more than its long legs and powerfully clawed feet.

Above
The blue peafowl originated in India and Sri Lanka but has been introduced into many parts of the world. The male's ornamental train takes several months to grow to its full size after the annual moult and is used in display.

Right
Instead of incubating its eggs, the mallee fowl builds a mound of earth and vegetation about 4 feet (1·2 metres) high and 15 feet (4·5 metres) across and lays its eggs in a central chamber. The decomposition of the 'compost' heats up the mound and the adults heap on or remove the covering soil in order to keep the temperature constant. When each chick hatches, it has to dig itself out and is not looked after by its parents.

12

specially shaped in order to assist the male bird in attracting a mate and defending his territory against others. In some species particular markings may be designed to startle or confuse a predator. A white rump or tail markings which suddenly become visible when the bird takes flight may cause a predator to strike for the most conspicuous but least vulnerable part of the bird. Alternatively or additionally, such markings may help a pair or a flock of birds to keep in contact.

Many birds are cryptically plumaged and this is especially important to incubating females which do not want to attract the attention of predators. Sometimes cryptic coloration involves a detailed imitation of the colour pattern of the surroundings. It may instead take the form of disruptive patterning–a striking design that breaks up the bird's outline. Counter shading is

very common, the bird being dark above and pale below. When the bird is on the ground this counteracts the natural 'bright above, shadow below' effect of light falling on a rounded shape and helps it melt into the background; in the air or in water, the bird presents a dark target against a dark background when seen from above, and light against light when seen from below.

Feathers also give birds the power of flight. Because of it they can exploit food resources not accessible to most mammals, travel long distances rapidly and evade predators. Relatively few mammals have the ability to exploit marine, arboreal or aerial situations in the range of ways that birds do. Also, mammals have not the speed of mobility to let them take advantage of seasonal flushes of food to the same extent that migratory birds can.

Right
A hunter of medium-sized birds, the peregrine falcon is found throughout much of the northern hemisphere, southern Africa, Australia and the tip of South America. The peregrine usually waits, flying high in the sky, for a suitable prey to fly past beneath it and then stoops down at high speed to strike its target with its feet.

Left
Projections inside the beak allow the puffin to catch and carry numbers of small fish at once. The webbed feet are used in surface swimming but underwater puffins and other auks use their short wings for propulsion. The brightly coloured outer casing of the beak is grown and worn only for the breeding season.

In flying, as in other activities, every technique must make good economic sense. In other words, the less energy a bird needs to use in finding food, the less food it needs to find. This makes its survival prospects greater when food is in short supply and also, by reducing the time that it needs to spend seeking food, reduces the time it is itself most vulnerable to predators. Thus, though birds have evolved a number of different ways of flying, all are governed by a rigorous need for efficiency.

Some groups of birds fly with minimum effort by utilizing natural air currents. When the sun warms the land surface, the air above it also warms and starts to rise by convection; the rising air currents are called thermals. Vultures, eagles and buzzards are amongst the birds whose long, broad wings can ride on the thermals and lift the bird in slow spirals high into the sky, often without any need for it even to flap. From this situation it can, for example, survey a large area for food. For long distance travelling, such birds can rise in one thermal, descend in a long glide across country and rise again in another. Thermals of this type are only created over land and they do not occur when the air is cold. Therefore, these birds cannot use this method of hunting in early morning or evening and, if they

are migrants, they usually cannot easily cross wide stretches of water but must travel round them. This is because the huge wing surface, so efficient for soaring, requires excessive muscular effort for sustained flapping flight and it cannot be maintained for long periods.

This is not to say that there are no air currents for seabirds to exploit. The albatrosses, famed long distance oceanic travellers, make their journeys by riding the air streams close to the surface of the ocean. Flying with the wind about 100 feet (30 metres) above the sea, the bird picks up speed and glides down towards the water. Here the wind speed is lower because of the friction between the moving air and the surface of the sea, so that the bird wheels round into the wind, is lifted by it, regains altitude and again turns down wind to repeat the process.

Birds which catch flying prey fall roughly into two groups–those which rely on high speed for interception in the open and those which develop manoeuvrability for hunting in cover. Falcons and swifts are in the first group and have long, pointed wings. Sparrowhawks are in the second group and have rounded wings which provide a large control surface acting against the air to facilitate weaving in and out through trees; they also have long tails which assist manoeuvrability

Left
Introduced to Australia, New
Zealand, South Africa and
North America, the starling
has been exceedingly
successful in some of these
countries where its
adaptability has allowed it to
exploit local conditions as
well as or better than the
native birds. It has become
a major pest in some areas.

Above
As it settles on to its nest,
the greenshank reveals both
its own cryptic plumage and
the camouflaged pattern of
its eggs. Breeding in marshy
areas across northern Eurasia
and wintering around milder
shores between Britain and
Australia, the greenshank's
long beak is used to probe
soft ground and mud for
insects and invertebrates.

and long legs to reach out and grab their prey.
What is good for the predator is also good for
prey species and similar basic wing and tail
shapes appear in other woodland birds which
have to move rapidly through trees.

Birds that hunt small prey that is hard to see
usually need to perch in order to scan for it. The
ability to hover allows kestrels, terns and others
to hunt where there are no perches. Superb
muscular coordination ensures that the bird's
head remains still so that its eyes can search
for prey, unblurred, however fast its wings beat
or its tail and body swing against the wind. These
birds hover by flying into the wind at the same
speed as it blows them back. Hummingbirds,
taking nectar from flowers on which they cannot
perch, are independent of air currents and hover
by making alternate forward and backward
strokes with the wings, so fast that their bodies
remain stationary.

A number of species which seek food under
water, such as auks and dippers, have evolved

Below
Its huge eyes adapted to making the best use of any available light and set on the front of its head to give good binocular vision for hunting, the collared scops owl is a widespread Far Eastern species. The brown plumage probably helps to conceal it from larger predators and mobbing small birds when it is roosting by day.

Above
In the breeding season, male ruffs don elaborate and individual breeding plumages and display communally at a lek to which the females come to be mated. The most dominant males will mate with the majority of the females which then depart to nest and rear the young unaided.

short wings for swimming beneath the surface. This imposes a penalty on their flying ability. They must beat their wings very fast and they find it difficult to gain altitude quickly. Auks may depend on the rising wind currents at cliff faces to help them ascend to their nesting places. Penguins have developed their swimming skill so highly that they have lost the ability to fly.

Between these various extremes there are many more general purpose wing shapes which are suitable for birds with less specialized life styles.

Flight imposes physical demands on birds. The most obvious reflections of this are in their skeletal structure. To produce a rigid airframe, able to withstand the stresses of flying, many parts of the body and limb bones are fused. Lightness has been obtained by the development of hollow bones—weight for weight both stronger than solid ones and providing a larger muscle attachment area. The flexibility necessary to feed, drink, preen and so on is provided mainly by the neck vertebrae.

Many specialized adaptations may be seen in the feet and beaks. All modern birds have evolved beaks which are lightweight, multipurpose tools. The grinding of food carried out in reptiles and other classes by teeth is undertaken by birds in an internal organ, the gizzard. This has a number of advantages. It allows birds to take in food rapidly, incidentally reducing the risk of predation; and it removes a heavy part of the bird's structure–teeth, jaws and associated musculature–from the mechanically disadvantageous position at the end of the airframe to a gizzard located inside it and close to the centre of gravity where it imposes least stress on the skeletal and muscular structure.

Beaks are immensely varied in size and shape to obtain and deal with different types of foods. Hooked shapes tear lumps of flesh from animal carcasses. Long, pointed bills provide an extended reach, perhaps to probe for nectar deep into flowers, to seek invertebrates in mud or to spear fish under water. Wide gapes facilitate the capture of insect prey in flight. Blunt, short beaks are best suited to coping with nuts and other fruits and seeds.

Feet have also evolved characteristics that assist the bird in its environment. Raptors usually have strong, sharply clawed toes to grasp their prey. Birds which walk on soft surfaces such as mud have long toes to spread the body weight. Swimmers have lobed or webbed feet. Perhaps one of the most important adaptations of all is the mechanism which makes a bird's toes lock on to a branch when it bends its legs into the roosting position. Thus many species can sleep securely in situations that are relatively inaccessible to predators.

Varied and valuable as these adaptations are it is flight that gives birds their special advantages, not least in being able to travel long distances to exploit temporarily favourable conditions.

Above
A female ruby-throated hummingbird hovers over her nestling. Hummers can hover in still air by beating their wings both backwards and forwards. The long beak is used to probe into flowers and they have long, tubular tongues to extract the nectar.

Above
Its beak crammed with insects, a wren returns to its domed nest in an outbuilding. Though vulnerable to hard weather, the wren is a widely distributed and successful species throughout Europe and part of North America, central and eastern Asia.

Migration

We know that migration has fascinated thoughtful observers of bird life for over 2 000 years. Even to those who sought no explanation for it, the return in spring of migrants such as the cranes, or the first calling of wild geese in the darkening skies of autumn, were to rural people both signals and symbols of the changing year. Today, though the start of the growing season and in its turn the beginning of hard weather are not obviously vital matters in the lives of our largely urban and industrialized society, we still welcome with pleasure the April cuckoo and swallows. As we begin to understand the methods by which they can navigate precisely across the world, and the reasons that make migration necessary, we must be still more fascinated and impressed by the complex and efficient biological machines that we call birds.

The purpose of migration is to move from one geographical area to another in order to exploit a habitat and food supply that is only seasonally available. There is a general southward shift of birds when autumn comes in the northern hemisphere. Many leave it entirely, moving into a new summer, far to the south. During the winter months, the lands around the Arctic are snow and frostbound so that few birds can feed there. Even as far away as the Mediterranean, food supplies for some birds, notably those that eat insects, are unobtainable or in short supply. Other species need travel less far, and Britain, which has relatively mild winters due to the effect

Below left
The sooty shearwater breeds in Tasmania, New Zealand, Tierra del Fuego and adjacent islands. It remains at sea outside the nesting season, following the prevailing wind systems to range far into the north Pacific and north Atlantic, feeding in the rich cold waters of both hemispheres in their respective summers.

Left
Ritual greeting between adult white storks at the nest. The European breeding population winters in Africa south of the Sahara. Storks make use of thermals to travel long distances so almost all the birds fly round the Mediterranean via the Straits of Gibraltar or the Bosphorus, where over 200 000 may pass through in autumn.

Below
The garganey is a small duck that breeds across Eurasia and winters well to the south, from Africa to Indonesia, favouring marshes, water meadows and small ponds with plenty of plant cover. These drakes are in Kenya and are accompanied by a Hottentot teal.

23

of the North Atlantic drift, holds large numbers of wildfowl and waders, as well as thrushes and other birds.

In spring, due to the rise in temperature and increased daylength, plant growth starts or accelerates again, insects and small mammals increase in numbers and the birds commence their northward movement, hurrying to breeding grounds where they can exploit the brief abundance of food that is required to rear their young.

The origins of the migratory habit are unknown. One theory is that it could have been developed or redeveloped towards the end of the last ice age. As the ice retreated each summer further than the summer before, species which learnt to move north after it would, as they do now, have been moving into territory where summer food was plentiful and competition from other birds reduced.

Migration places great demands on the birds which undertake it. Their timing, fitness and navigational skills must all be top class if they are to succeed and to survive. Even before it is time to leave their winter quarters, the lengthening daylight and increasing temperatures trigger off physiological mechanisms that prepare the birds. Those long-range migrants which make their whole journeys almost non-stop accumulate large quantities of fat; some may more than double their total body weight at this time. The fat is the fuel which they will burn up in flying con-

tinuously for perhaps 1 000 miles or more and remaining airborne for many hours. Of course, birds which travel in more leisurely style, stopping fairly often to feed en route, do not need to build up fuel reserves to anything like this extent.

The timing of departure is important. Birds that arrive too soon at their northern breeding grounds, when the weather is still harsh, may starve or freeze. Those that arrive too late will find the best nesting territories already occupied and their chances of breeding successfully will be reduced. Often males arrive shortly before females so as to be holding territories as soon as possible. The return timetable seems much less critical and many adults leave early, while food is still plentiful. This must leave more food available for the young birds preparing for their first great journey and also avoid any risk of the whole population being caught and destroyed by a freak early cold spell.

Some species migrate by night, others by day and some are not governed by light conditions. Most of the small birds that feed on insects move after dark because this leaves the day free for them to search for food and lessens the likelihood of predation. Those that feed on the wing, such as hirundines, can eat as they go and so migrate by day as do the majority of small seed-eaters. Birds of prey are diurnal migrants; broad-winged species such as buzzards, which progress by soaring up on one thermal and gliding down to

the next, must wait until the sun warms up the land surface before travelling. Many waders and waterfowl move by either day or night; the feeding rhythm of those that live along the shore is governed by the tides rather than light and darkness.

Birds whose migratory movements are largely overland generally move on broad fronts but there is a tendency for them, when they reach a coastline, major valley or mountain range, to follow it for as long as it runs roughly the right way. Only where it turns from the intended direction of migration do they strike out on the crossing. Thus headlands on the south coast of Britain are often good places for seeing birds on migration. Birds which fly using thermals must remain over land and skirt round large water bodies. Vast numbers of birds of prey converge at land bridges such as the Bosphorus between the expanses of the Black Sea and the Mediterranean. These passages are particularly spectacular in autumn when numbers are swollen by the young of the year.

Radar observation has shown that many birds migrating at night fly at around 3000 feet (914 metres) and most below 5000 feet (1524 metres), though in cloudy weather they may fly much lower. However, some species habitually fly higher than this, notably where they must cross mountain ranges. Bar-headed geese cross the Himalayas at about 30000 feet (9144 metres) in air temperatures well below freezing point.

Travelling speed varies considerably. The overall rate is often fairly low because the birds make stops to rest and refuel. However, some can certainly travel rapidly for remarkably long distances. The eastern race of the American golden plover is known to move from its Nova Scotia breeding ground to the wintering area in South America, a distance of 2400 miles (3862 kilometres), in about 48 hours which requires flight at an average speed of 50 miles per hour (80 kilometres per hour) for two days and nights. It is believed that sedge warblers may be able to travel as much as 1500 miles (2413 kilometres) non-stop at about 25 miles per hour (40 kilometres per hour)–that is 60 or more hours flying–and thus move in one stage from England to south of the Sahara.

Many birds migrate in flocks. Geese and cranes are obvious examples which may be seen moving by day in long skeins. This formation flying, which can also be seen in gulls, waders and some other birds, has a functional advantage. The slipstream set up by each bird in line helps the one behind it by providing extra lift and thus reduces the energy expenditure. Periodically the leading bird of all, which does not get this benefit, falls back and allows another to take over. It does not seem that smaller birds fly in this way, perhaps because they do not generate enough slipstream to make it worthwhile, though of course movement in a flock may give the individual some protection from predation. When moving at night, flocking species apparently call to each other so as to keep in contact. The calls of redwing are common in British skies on autumn nights.

Wind and other weather conditions are very important factors in the timing of migration and

in its success. In spring, the onset of warmer weather, and in autumn, a change to cooler weather, generally starts the main movement off but beyond that birds choose to travel when it is not raining, when skies are clear and when winds are light. The timing is particularly important for birds which undertake sea crossings. If head winds spring up after they are well on their way or if cloud or fog mist closes in and they cannot navigate successfully they may be forced off course and land somewhere far from their normal route. If they are unlucky, they may not be able to make landfall before they run out of fuel, sink exhausted to the sea and drown. There is no doubt that migration is costly. Every year many birds do not survive it. But the benefits of successful migration to a place where food is reasonably abundant are so great as to make it worthwhile for a species to evolve this life style.

The greatest question about migration is how do birds find their way? They may travel across thousands of miles of featureless ocean and yet return not only to a general area but to some small and remote island in the Pacific, or to one particular, tiny lochan out of many hundreds in the Shetlands, or to the same Arctic nest site that they used last year. Also, journeys are often made with no guidance from those who have gone before; in many species the adults leave the breeding grounds before the young, which have to find their own way to the wintering area. If they belong to a species which does not reach maturity in its first year, these young birds may not retrace their flight until more than another twelve months have elapsed. Complicating matters still further, birds of the same species may not all be moving to the same general area. You may see dunlin that breed in Britain, already in their summer plumage, mingling on a south

Above
The hobby feeds primarily on insects caught on the wing but is fast enough to take birds such as sand martin and even swifts. It favours fairly open country provided there are tall trees for nesting nearby. Because their food becomes scarce in winter, hobbies migrate south into southern Africa, India and south-east Asia.

Left
Short-eared owls range over much of Eurasia, America and other areas. Hunting prey in open country, usually small mammals, the birds disperse widely at times of shortage, unlike woodland species which must remain in their familiar territories if they are to find food.

coast estuary with other dunlin still in winter dress bound ultimately for breeding grounds that stretch for thousands of miles across the tundra.

Many experiments have been carried out which demonstrate the ability of birds to find their way home across unknown terrain. Some of these studies are amongst the classics of ornithological research. In 1940, 220 Leach's petrels were taken during the breeding season from their colony on an island near Nova Scotia and released at points up to 470 miles (756 kilometres) away. Most of them returned easily. A still more impressive demonstration of the ability to return home from a distant and unfamiliar area was provided in 1968 by individuals of the same species brought to England. They returned the 2 918 miles (4 684 kilometres) in less than 14 days, travelling at an average speed of 217 miles (349 kilometres) a day. Similarly, a Manx shearwater taken from the island of Skokholm off the Welsh coast and released in Boston, Massachusetts returned in

12·5 days, crossing the 3 200 miles (5 149 kilometres) of sea at a rate of over 250 miles (400 kilometres) a day: it arrived before the letter informing the researcher that the bird had been released. Of eighteen Laysan albatrosses taken from their nesting area on Midway Atoll in the Pacific to six different locations, fourteen returned, including individuals from all six locations, the furthest of which was 4 120 miles (6 629 kilometres) off in the Philippines – a distance covered in 32 days. The fastest rate of return was over the 3 200 miles (5 149 kilometres) from Washington in 10 1 days at an average speed of 317 miles (510 kilometres) a day.

Not only seabirds have this ability. Displacement experiments with purple martins in the USA included a bird which flew back to its nest, a distance of 234 miles (376 kilometres), overnight in 8·58 hours.

A full explanation of how birds navigate still remains to be worked out but experiments with birds in the field and in artificially controlled environments do provide some indication of the range of their abilities.

It has been proved that the young of some species have an instinctive knowledge of the direction in which they should fly in autumn. In one experiment, starlings were caught in Holland as they moved south-west to winter in southern Britain and north-west France. They were taken to Switzerland, about 400 miles (644 kilometres) to the south, ringed and released. The young birds, which had never before made the journey to their wintering area, continued to fly in a south-westerly direction and thus largely ended up in an area centred about 400 miles (644 kilometres) south of their proper wintering place. In the subsequent autumn, those birds that had been displaced as juveniles returned again to winter in this wrong area that they had reached in the first year. However, adult birds similarly displaced made no mistakes but at once corrected their route and flew north-west to return to the area they had used in previous winters. This suggests that in the case of starlings, young birds make their first migration on the basis of an instinctive knowledge of the heading on which

they should fly and the distance they should go. Once they have made the journey they know the place to which they should return and have the ability to find it even when displaced to an unknown location. What are these navigational skills and do we understand how these skills function?

Experiments with caged migrants have shown that species which normally travel by day will flutter against the side of the cage facing towards their normal direction of migration, provided they can see the sun. If the sun's apparent position is changed–for instance by using mirrors–the birds are deceived and will reorientate their position in the cage. When the sun is obscured altogether, they cannot orientate themselves correctly. Similar experiments with nocturnal migrants show that they navigate by the stars and they too are lost when the stars are obscured.

The principle of solar and stellar navigation is simple. For example, at noon the sun is due south so by setting course towards it one is travelling south. The sun moves round so that by 3 p.m. it is in the south-west, at 6 p.m. due west, at 6 a.m. due east and so on. By making

Left
Three sanderling stand with several dunlin. Dunlin nest widely on northern moorland and tundra but sanderling are confined to the high Arctic and their entire breeding season lasts only six weeks. Both species winter in coastal areas—dunlin concentrated in estuaries in the northern hemisphere and sanderling spread along shores all over the world.

Above
Common cranes breed in marshy areas of northern Europe and Asia, and in winter move south in flocks. Unlike storks, they do not use thermals and often travel in V-formation as the spreading air eddies set up by each bird help the one behind to fly with less effort.

accurate allowance for its movement with the passage of time, course corrections can be made and a straight direction of travel maintained. Stars move similarly by night. For man, the sun and stars were for long the only means of navigation at sea but such navigation was never very accurate on long voyages out of sight of landmarks until clocks were developed that would measure precisely the passage of time.

Migrant birds must have an internal 'clock' on which they can base course corrections. Those which, like the sedge warbler and the American golden plover, migrate non-stop day and night, must be able to use both the sun and the constellations as guides for navigation.

Accepting that young birds know the direction in which they should fly on their first migration, how do they know when to stop? It has been observed that captured migrants become restless with the onset of the time when they should be moving off on passage and that after a period this drive abates. We therefore assume that the urge lasts long enough for the young birds to make the journey to the right wintering area and that it then stops, so that the bird settles in the appropriate place.

Apparently if a bird has been to a place once, it can, as we saw with displaced starlings, move back to it even when it finds itself somewhere it has never been to before. This is an important ability for birds blown off course by storms. Again the method is simple in theory. If the bird is

Left
Adult and young house martins gather in excited flocks before the start of their migration. Feeding on aerial insects as they go, they travel only by day. Reedbeds are an important roosting habitat for many hirundines, though house martins may also use other sites.

Right
A swallow approaching its nest, its throat distended with insects. European swallows move into Africa in winter. When they return, birds that breed in southern Europe arrive first and may already have flying young before those that nest in the extreme north pass through.

moved to the east, this has the same effect so far as the sun's position in relation to the bird is concerned as if the sun itself had travelled on westwards–that is, as if the time of day was later. But the bird's internal 'clock' knows what time it is and so it strives to put the sun back in the correct position in relation to the time by moving itself to the west. Similarly, displacement to the south brings the observed sun further north and thus to a higher position in the sky than it should be at that season, so that the bird moves north to put the sun back in the position that the bird expects it to be in. These correcting movements combined finally lead the bird back to where it was before displacement.

However, solar and stellar guidance cannot take a house martin back to the very eaves under which it nested last summer or a jack snipe back to the little marsh in which it passed the preceding winter. Such precise navigation clearly depends on visual knowledge of the locality.

Studies with homing pigeons have shown that they know the landmarks for about 10–12 miles (16–19 kilometres) from their home–that is an area of over 300 square miles (768 square kilometres)–and can readily locate their lofts. Such a homing ability is valuable for at least two reasons. If a bird is successful in a given site in one year it is likely to be so in the next and thus it is useful to be able to return to such a known site rather than to have to search afresh for one each time, when failure will mean either that the bird cannot nest or that the winter may kill it. The other advantage is a genetic one. By returning to the same limited area each summer, a group of birds becomes a gene pool in which desirable evolutionary traits best suited to that area are retained and reinforced instead of being dissipated and lost as they would be in a very fluid population.

It remains to be discovered what further navigational skills birds may have. Not all lose their sense of direction when clouds obscure the

Left
The black swan is an Australian species. Feeding on aquatic vegetation, food shortages caused by periodic droughts make it necessary for the population to be highly mobile and the species has no set breeding season but will nest whenever conditions become suitable.

Right
The dark-faced ground tyrant is a ground-feeding flycatcher which nests largely in the most southerly part of the Andes in Chile. The ten other Andean species in the family nest at different latitudes along the range, choosing different altitudes to obtain similar climatic conditions. In winter all move north.

sky and it is likely that they use other clues to help their navigation at these times. A wind blowing steadily from one direction could serve as a reference point and a bird's senses of smell and hearing may also be brought to bear.

In considering migration, some attention must be given to other periodic movements by birds. A common variant of normal long-distance migration may occur in montane species such as the Himalayan monal which moves up to higher altitudes in summer and down to the foothills and valleys in winter. Some wildfowl such as shelduck carry out a 'moult migration'. They become flightless during moult, and so move before its onset to the same site which has always in the past proved to be safe and to hold adequate food for their survival during this exceptionally vulnerable period.

Particular weather conditions may also cause substantial movement. Exceptional hard weather in central and eastern Europe forces wintering birds west and south to seek milder conditions. Birds which live in arid areas may be dependent on rains for breeding and will move accordingly. For example, in Africa the quelea undertakes mass movements to take advantage of the irregular rains and the consequent abundance of seeds on which it feeds. In Australia, where seasonal temperature extremes are much less than in countries where normal migration is a feature, birds do not move on a regular seasonal basis. However, when the rains do occur in the arid areas there are massive immigrations of birds into the suddenly hospitable area – predatory species following the seed-eaters and insectivores, and all raising young while food is plentiful.

Periodic movements of birds out of their normal ranges also occur and are particularly a feature of birds of the boreal forests. These movements, which are called irruptions, may be related to fluctuations in the food supply. It is well known that when the population of that small Arctic rodent, the lemming, builds up to a high level, vast numbers move away from their normal areas and the population crashes because many die. This occurs every few years on a fairly regular basis. It affects those birds such as the snowy owl and rough-legged buzzard which feed on lemmings. When their populations are no longer sustained by the food supply, they too irrupt widely and turn up far outside their normal ranges.

Migration is not an end in itself but a means to an end. It allows maximum exploitation of world-wide habitats and food resources by bird life. To see how birds relate to their habitats and to each other within them, it is necessary to turn our attention to the structure and opportunities that each habitat offers.

Left
The mallard is a widespread and successful species breeding in tundra, boreal, temperate and steppe habitats. In many areas it is resident but in part of its range it moves south in winter. Young birds may disperse widely before settling to breed.

Above
Razorbills breed around the North Atlantic. Coming ashore only to nest, the birds spend most of their lives on the sea. Ringing has shown that the north European population moves south along the continental shelf in the winter to waters off Spain and North Africa and in the Mediterranean.

Above
The nutcracker is a boreal
forest member of the crow
family. Feeding mainly on
conifer seeds, its population
fluctuates with the cone
crop, a bad season following
several good ones forcing
the birds that cannot find
food to move south far
outside their normal range.

Polar Regions

The polar ice-caps, the cold, stormy seas around them and the vast, treeless tundra would seem to be the most testing and hostile environments that birds could encounter. Certainly the number of different species found in these areas is comparatively small. However, this is because they are relatively simple environments, particularly when compared with terrestrial habitats in temperate and tropical regions, so that the number of ecological niches in which different species can develop different ways of living is relatively small. The species that do live here have been remarkably successful in overcoming the climatic problems and some are extremely numerous.

The Antarctic region covers about 18 million square miles (45 million square kilometres)–that is an eleventh of the whole earth's surface. At its centre lies a continental land mass, most of which is buried deep beneath the ice. Around the icecap lies a great ocean stretching north to the southern tips of Africa, Australia and America, sprinkled unevenly with islands and partly covered by pack-ice in winter. In mid-winter, this

Left
An adult Adelie penguin with its two chicks, both covered in dense, warm down. When the young become too big to be brooded, they gather in creches where they can huddle together for warmth and are less likely to be attacked by predators such as sheathbills and great skuas.

zone receives little daylight, but as summer approaches the days begin to lengthen, the temperature rises and the pack-ice retreats. By mid-summer, with daylight almost continuous around the clock, ground temperatures have risen considerably and a partial thaw occurs on the coast. The continent is very dry, with remarkably little precipitation either as snow or rain and so very little water to support plant growth and provide food. Winds are frequent, in some areas almost continuous and often violent. They cause warm-blooded organisms to chill much more rapidly than they would in cold, still air and thus increase the problem of keeping warm. They also blow fallen snow back into the air and create blizzards and drifts.

For land birds to find food in this continent and on the islands near it is enormously difficult, and only five species do. The remaining thirty-eight species that breed in the region of the high Antarctic are all seabirds.

The first problem faced by any bird which penetrates these regions is the cold. Seabirds have a significant advantage in this respect because they are already, to some extent, equipped to deal with low temperatures. Even in tropical seas, the water temperature is below body heat and so all birds which regularly enter water have to be insulated. They have waterproof plumage and beneath the skin a layer of fat which helps to prevent heat loss. They tend to be relatively large and compact compared to many land birds and this gives them a low proportion of surface area relative to their body size so that they radiate heat less rapidly than smaller species do. Thus birds capable of dealing with marine situations are already partly suited for polar climates.

The Antarctic seas are rich in food. Warm currents from the north and cold ones from the

Right
The emperor penguins are amongst the world's biggest seabirds. Their streamlined shape and powerful flippers make them most efficient swimmers, usually well able to outswim the leopard seals that lie in wait at the edge of the pack ice. Though larger than the king penguin, they have smaller extremities to reduce heat loss.

Above left
Bewick's swans breed across the tundra of Asia and America, the same pair of birds returning to the same nest site each year. Before flying south for the winter, the adults moult their flight feathers and remain on or near open water until the new set has grown.

Below left
A cock willow grouse in autumn plumage. Like ptarmigan these birds have a white winter plumage and are brown in summer. They have a wider habitat tolerance, occurring not only on tundra but also in birch woods, conifer or willow scrub and on heather moors. The Scottish red grouse is a subspecies which does not turn white in winter.

south meet to produce conditions ideal for seabirds. Plankton and fish are extremely plentiful, and birds, seals, squid and whales compete for them.

Albatrosses, shearwaters and petrels are by far the most numerous Antarctic birds and occupy a range of niches with different feeding and nesting requirements. At the one extreme come four albatross species which breed on islands in Antarctic or sub-Antarctic water. Spending most of their lives at sea, their contact with the region can be limited to summer exploitation of breeding places and the rich supplies of fish and squid. By comparison, many shearwater and petrel species spend their whole lives in these waters.

The giant petrel, which has a wingspan of 6·5 feet (2 metres), almost the size of an albatross, is a major predator of smaller birds, eats carrion such as dead seals and whales, and also hunts fish and squid. The various shearwaters and smaller petrels take fish or other marine food including plankton and have distinctive feeding habitats by which they minimize inter-specific competition. For example, Wilson's petrels flutter and hover at the surface to feed on plankton, while diving petrels hunt small crustaceans below the surface.

Best adapted of all birds for the extreme rigours of Antarctica are some of the penguins. All penguins are flightless and when they are not involved in breeding and so confined to land, they spend their time in the water. They therefore have to be particularly well insulated. Contour feathers cover the entire body surface excluding beak, feet, eyes and brood patch at a density of about 75 per square inch (12 per square centimetre), overlapping each other and lined with down. The thick skin is underlain by blubber. Feet and flippers, which are not insulated, are kept at low operating temperatures. Warm blood is pumped outward from the heart and flows through blood vessels that run close to others carrying colder blood back from the extremities. Thus the warm blood is cooled and the cool blood warmed, with the result that body heat circulates within the body core, the blood carrying oxygen to the extremities being pre-cooled so that heat loss is minimized. Because of these adaptations, penguins are in danger of overheating when very active and at such times the birds pant, the blood vessels in the extremities are dilated and the feathers are

Top
The Antarctic tern nests in small colonies on islands in and around the Antarctic ocean. As with other terns, all members of a colony cooperate to defend the site against intruders such as marauding gulls and skuas. Birds from the southern part of the range move north in winter when the sea freezes.

Above
The black-throated diver breeds on inland waters in northern North America, Asia and Europe. The nest is built near the water's edge as their legs are set so far back on the body—for maximum swimming efficiency—that they can hardly walk. In winter the birds move to unfrozen coastal waters.

Left
A black-billed sheathbill with gentoo and king penguins. One of the few non-seabird species to live in the Antarctic, they are scavengers particularly in penguin colonies where they steal regurgitated food brought to the young. They will also take eggs, kill sick or injured birds and eat shellfish.

Below
Perhaps the most numerous bird in the world, Wilson's petrels breed in vast numbers in Antarctica. Like most other petrels they nest in holes or rock crevices and usually come ashore only at night to avoid predators. They patter over the sea and dip their beaks to feed on plankton.

raised to allow the increased escape of warmth.

The species which lives under the hardest conditions is the emperor penguin. Standing 4 feet (1·2 metres) tall, they have to rear correspondingly large chicks and this takes time and much food, especially as they become larger. Thus, in order to complete rearing the young during the summer when food is plentiful, the adults start their breeding cycle before the onset of the preceding winter. At this time they move south towards their regular breeding sites, walking or sliding over the ice for up to 60 miles (96 kilometres) or more inland. The birds mate and eggs are laid about June. Within a few hours of laying, each male takes the egg from his mate, balancing it on his feet, and draping a feathered fold of skin over it so that it is tucked warmly against the bare brood patch. As the female has now been without food for several weeks and has had to produce the egg from her own stored food reserves, she needs to feed. She abandons the male and heads north for the sea. He is then left to incubate the egg without respite for two months. This means that, including the courtship period, the male will go without food for about four months, living on reserves in his body. Breeding communally, the incubating males crowd close together so that only the smallest possible part of each individual is exposed to the elements. This helps them to keep warm and so

they burn up their fuel reserves much more slowly than they would otherwise. Experimental studies have shown that an emperor penguin kept on its own in the same weather conditions will lose weight two or three times faster than one in a crowd. Thus the communal breeding has survival value and is vital for success.

The females return at about hatching time. If the chick is born before its mother arrives with food, the male can still produce a rich secretion on which to feed it. Now the females take over and the males return to the sea to break their long fast and then return with more food. By the time it is about two months old, the chick can stand up and, covered with a thick downy coat, is left in a dense pack with other juveniles while both its parents fetch food. It now receives about one-third of its body weight in food every two or three days and the parents' journey to the sea becomes progressively shorter as summer advances and the sea ice retreats. About mid-summer the half-grown young birds begin to fend for themselves, travelling to sea with their parents.

It is not yet known whether the emperor adults can breed every year or whether they need twelve months to recover. Although the problems would seem great, breeding success is fairly high. By mastering conditions that few other birds can endure, emperors diminish losses to predators and by starting the breeding cycle at the hardest time of year, the young are ready for sea when the

weather is at its most mild and conditions for them are easiest.

The king penguin is also a large species, standing about 3 feet (0·9 metre) high, and it faces similar problems in rearing a large chick, but has solved them by quite a different method. Kings nest in colonies close to the sea, mating in late spring so that the single chick is hatched just after mid-summer. Because of their proximity to the food supply, feeding proceeds rapidly and by autumn the chicks are well-grown. Then, because food is becoming scarce with the onset of winter, the adults abandon them except for a feeding visit from one or other parent about every two or three weeks. In the same way that male emperors do, king penguin chicks huddle together to minimize heat loss and thus lose weight only slowly, but even so the smaller individuals die. When weather and feeding stocks improve again in spring, the adults can supply the chicks more often and they once again begin to gain weight, reaching independence that summer.

Adults whose chicks die in the winter will be able to breed early the following season so that their next chick, having grown large by the autumn, will be more likely to survive the following winter. Each pair of birds normally manages to rear a chick in every second or third year and this recruitment rate is sufficient to maintain the population.

Other Antarctic penguin species are smaller and can rear their young in a single summer. Those which breed on the continental coast as well as the islands, such as Adelies, are also well insulated and their blubber layer further serves to protect them from physical damage as they leap ashore on to the ice or rocks from the sea. Species such as rockhopper and Macaroni which breed only on the sub-Antarctic islands are less equipped for polar conditions but by nesting amongst rocks they exploit the most sunny and sheltered positions available.

In contrast to the Antarctic, which is a land mass surrounded by water, the Arctic is a water mass largely surrounded by land and this makes significant differences in its nature and in the opportunities provided for birds. The land areas are seasonally available to large numbers of migrants which winter in temperate and tropical regions. The marine situation is similar to that in the far south. The seas are very fertile where cold and warm currents meet and mix their nutrients and oxygen, and they support enormous quantities of plankton.

There are no flightless seabirds in the Arctic. The one flightless species which did exist in recent times, the great auk, was over-exploited for food by sailors and became extinct in the first half of the nineteenth century. There are more gulls but far fewer petrel species than in the Antarctic—only three breeding species as against two dozen. The niches occupied by penguins and some petrels are largely taken in the Arctic by species of auk and some gulls.

In examining the respective seabird populations, it is interesting to compare the common diving petrel of the southern oceans with the little auk. The two species are unrelated and live literally at opposite ends of the earth. Yet, because they hunt similar food in similar ways, they have evolved similarity of appearance and structure. Both feed on plankton caught under water. Both use their rather short wings for

Above left
A flock of snow buntings searching for seeds in the snow. Relatively few small birds can tolerate Arctic conditions but this species breeds throughout the tundra and in tundra-like habitat as far south as the Scottish Highlands, moving mostly to snow-free coastal areas in the winter months.

Above
Long-tailed skuas are thinly distributed throughout the tundra zone in summer where they feed primarily on lemmings, though berries and fish offal are also taken when available. Their wintering movements are not fully known but many birds are seen in mid-Atlantic in late summer, off the Argentine in November and around the Caribbean in early spring.

submarine propulsion. They have a whirring flight, usually low over the surface. Both enter the water by flying straight into it and leave it in what appears to be an instant transition from submarine swimming to rapid flight. Both are black and white – a common camouflage effect for marine birds – and have short beaks and tails, similar sizes and skeletal structures. This process whereby unrelated birds have developed similar life-styles and body features to deal with similar situations is called convergent evolution. It is a response over a long period of time by the two species to the exploitation of a similar situation in different areas.

Only the Dominican gull penetrates deep into the Antarctic but a number range into or are confined to the Arctic region. Amongst the latter is the large glaucous gull which has a circumpolar and largely coastal distribution and is a major predator on other birds, particularly little auks. Occupying a similar niche to the greater black-backed, the two species compete where their ranges overlap but generally they are geographically separated because they favour different climatic conditions. The smaller Iceland gull has a more limited distribution than the glaucous gull and feeds more extensively on fish. It too has counterparts in warmer latitudes and like them it avoids competition with the large gull species that share its range by exploiting different food.

Where the polar effects extend over the land masses, there are huge tundra areas that are ice-free in summer but where the growing season is too short or too severe to support trees. This zone stretches right round the far north of Canada, Alaska and Siberia, through Iceland and around the coasts of Greenland. In winter the snow is frequent, day length is very short and the ground is frozen. In summer, the surface thaws and, because the sub-soil remains frozen and drainage is impeded, many lakes, pools and marshes form. The daylight is long and there is a rich growth of ground vegetation with dwarf willows and birches in sheltered areas. Insect life is very plentiful. For this short, rich season, the tundra becomes alive with birds – notably divers, waders and wildfowl – most of which are long range migrants from the south.

Because there are no trees, tundra breeders nest on the ground though some species, such as barnacle geese, choose cliff ledges and offshore islands where they are safe at least from Arctic foxes. As ground vegetation is often very short and gives little cover, the eggs and females of tundra breeding species are almost invariably camouflaged. A bird which may appear strikingly marked and very noticeable when standing on an English estuary in winter can vanish easily when it sits down on its nest in the Arctic tundra.

Above
The huge snowy owl is resident throughout the tundra zone, only moving south when the lemming population collapses, as it does about every fourth year. As in most birds of prey, the young do not all hatch at once, so that if food runs short the largest chicks have a good chance of survival at the expense of the smaller ones.

The tundra breeding season is short and it is important that no time is wasted. Thus many birds pair up before they leave their wintering grounds–ducks begin their courtship in winter because of this, and in geese and swans it is known that the same two birds will return each year to the same nest site thereby avoiding any delay in finding mates and nesting locations. Despite these precautions, tundra species can suffer large-scale breeding failures. For example, dark-bellied brent geese which breed in Arctic Siberia may fail altogether if the thaw is late and they cannot start the nesting cycle in time to finish before winter closes in again. There may also be other causes of failure–storms causing flooding or an early onset of cold weather. As a result, the populations of many wildfowl and wader species fluctuate considerably from year to year–especially as several bad years or several good years may come in succession.

Predation is a constant hazard here as everywhere. Gulls and skuas are not merely opportunists who will grab unprotected young, but will deliberately set out to harrass the adults until they can create a chance to seize eggs or chicks. Wildfowl in particular produce very large clutches because their losses are so high and there is no time to produce a second brood if the first fails.

There are also a number of land birds which are major predators. The snowy owl with its 5 foot (1·5 metre) wingspan is fully capable of surviving the Arctic winter. It feeds mainly on small mammals, notably lemmings, but also takes birds and other live prey in summer and Arctic hares and ptarmigan in winter, when most other birds are absent and lemmings are in hibernation. Like other species which feed on prey whose numbers fluctuate greatly, snowy owls react quickly to an increased lemming population, laying much larger clutches–perhaps as many as nine eggs instead of four or five. When the lemming population crashes, as it does every few years, the increased owl population is left short of food and many are forced to move south.

Snowy owls are well adapted to their Arctic existence. Their largely white plumage must have some camouflage value in hunting but is probably more important for heat conservation as the cells of white feathers (and fur) are air filled and this improves their insulative qualities. The plumage is very dense, with thick down, and the legs and feet are largely covered by feathers. In summer, after moulting, the thickness of the plumage is reduced.

Like the snowy owl, the ptarmigan on which it preys finds winter survival value in a dense white plumage. The ptarmigan moults in order to change its coloration to suit the prevailing conditions. White-feathered body and legs in winter change in spring to a dark brown, finely barred with a lighter golden brown. Its wings and underparts remain white but these are not apparent so long as the bird is sitting on a nest or walking, which it prefers to do rather than flying. In late summer, it adopts a duller plumage. While in moult, both going into white winter coloration and when coming out of it, the birds have a mottled brown and white coat which is particularly effective when a sprinkling of snow is on the ground.

Ptarmigan live mainly on vegetation, usually the shoots of heather, bilberry and other plants. These foods are present throughout the year and though much may be buried by winter snows, sufficient remains accessible particularly in more windswept areas for these birds to remain in the tundra zone throughout the year. In extreme weather conditions, ptarmigan dig out holes in the snow in which they can shelter away from the elements, insulated by the surrounding snow

and well camouflaged by their white plumage.

There are a handful of small birds that manage to live in Arctic conditions at least during the summer months. Snow buntings breed throughout the tundra zone. Males return first to the breeding area to take up territories when snow may still be on the ground. Thus, the females need not return until conditions are suitable for them to select the nest site and get straight down to breeding. Because the males have already selected territory, none of the short summer is wasted. They feed on insects during this season but in their winter quarters on temperate coasts they mainly eat seeds.

Tundra conditions are not found exclusively round the Arctic and–on a vastly smaller scale–the Antarctic. Small areas at high altitudes in other latitudes, such as the Cairngorm range in Scotland, also have tundra-like conditions and they hold species, such as ptarmigan, snow

Above left
The little auk is about the size of a starling and is one of the most numerous birds of the high Arctic seas, nesting in huge colonies in Greenland, Spitsbergen and the Russian islands. During the breeding season many birds prey on them, including glaucous gulls, Arctic skuas and gyrfalcons.

Below left
Northern giant petrels fight over the carcass of a seal. This species, breeding north of the Antarctic convergence, is usually darkly plumaged but the southern giant petrel tends to be white. Both are predators, killing gulls, other petrels and penguin chicks as well as acting as scavengers.

Above
Sabine's gull breeds at scattered locations on islets throughout the high tundra zone. The young are fed on insects and leave the nest as soon as they are hatched; this reduces the likelihood of their detection by predators.

Right
There are three populations
of barnacle geese. Ringing
studies have shown that one
breeds in Greenland and
winters in Ireland and Islay;
one breeds in Spitsbergen
and winters on the Solway;
the third group breeds in
Russia and winters in
Holland. Thus they could
eventually become distinct
subspecies or new species.

Below
The white-crowned sparrow
feeds on insects, berries and
seeds, breeding across the
high tundra in America and
down the western side as far
as California, having
probably gradually extended
its range down the mountain
chain into a progressively
warmer zone.

bunting and breeding waders, which also occur in true tundra habitats.

In the past, man's impact on the polar regions has been as a hunter. Eskimos in the Arctic killed to provide for their own needs. Later other nations hunted the mammals, notably whales and seals, for oil and furs. Seabirds were taken to provide fresh meat on voyages and thus the great auk was exterminated. The exploitation of the Antarctic animals did not begin until the late eighteenth century and though the brunt fell on seals and whales, birds too suffered. Vast numbers of king penguins were killed for their oil during the nineteenth century and they were exterminated from many breeding localities which they are now recolonizing.

Having reduced the seal populations and brought the larger whales to the brink of extinction, the modern Antarctic hunters are turning their attention to the krill—the larger plankton on which seals, whales and many of the seabirds depend. The resource is certainly an enormous one but judging by the past and present record of man's ingenuity and greed, one can have no confidence that it will not be ruthlessly and grossly over-exploited.

So far the habitats on which these creatures depend have suffered little, but now the world's oil comes not mainly from birds and whales but from petroleum, and the American environmental movement has lost its fight against the trans-Alaskan pipeline despite the great hazard that this poses to the fragile tundra ecosystem. Mineral exploitation of the polar areas will undoubtedly increase as supplies in more rapidly accessible regions of the world are exhausted.

Boreal Forests

South of the tundra lies a belt, sometimes narrow, but often hundreds of miles wide, where willow and birch, larch and spruce survive mostly in stunted forms or thinly scattered and only growing well in isolated sheltered pockets. This is a transitional zone between open tundra and the dense coniferous forest to the south where the trees grow thickly and continuously in a belt which stretches for about 8000 miles (12870 kilometres) across the north of Europe, Asia and America.

Summers are short and in much of the area the sub-soil remains frozen all year round, only the surface layer thawing. As in the tundra, drainage is poor and there are many swamps and lakes. In winter, it can be extremely cold with temperatures falling below minus 45°C. Rainfall is low and for most of the year the frozen ground makes it impossible for trees to take up moisture so the growing season is short. Coniferous trees are specially fitted to live in this environment and firs and pines are dominant. Their thin, needle-like leaves resist desiccation in extreme winter cold or drought. Most species retain leaves

Left
Goldeneye breed throughout the northern forests, building their nests in the cavities in trees growing near water. The same nest may be used for many years in succession. Soon after hatching, the young clamber out of the hole, drop anything up to 60 feet (18 metres) to the ground and take to the water as quickly as possible.

Right
The magnolia warbler in winter plumage. This is one of the most numerous warblers of the Canadian and Alaskan forest. They take insects and spiders in summer but many small berries are eaten in their wintering range, which is mostly in the southern and coastal USA and the West Indies.

Below
The Siberian jay is a largely sedentary species in boreal Eurasia. A related species, the grey jay, occurs in North America and the Szechwan grey jay is confined to the conifer forest in the mountains on the Tibetan-Chinese border, having probably been isolated there since at least the last ice age.

throughout the year and so they do not lose valuable and slowly-replaced nutrients by shedding them in autumn as deciduous trees do.

In the main forest zone, the trees grow close together and what little light there is for much of the year finds it hard to penetrate to the forest floor. Partly because of this and partly because the fallen conifer needles decay only slowly and carpet the ground between the trees, there is little or no ground vegetation. In winter, the snows may form a blanket over the tree-tops and the forests are then more sheltered from frosts than the open swamp areas, though the latter become warmer in summer than the shady and cool forest interior.

In the south of the zone, there is an uneven transition to deciduous woodlands. Conifers continue to thrive at high altitudes or on poor soils which, though they may be well to the south of the boreal belt, have similar growing conditions—for instance, in the Caucasus alongside the Black Sea.

Food resources in the forest are surprisingly varied. The conifers themselves are the primary source. Their bark, buds and seeds provide food for insects, mammals and some birds. Even the wood itself is eaten by the larvae of some insects. Many birds feed on the insects and in turn higher predators hunt other birds and mammals.

As in the tundra, many species occur throughout the whole boreal zone because the entire habitat is very similar. Often, several sub-species will exist. Living in different regions, each is slightly distinct in one or more respects

from those alongside it but interbreeding may
take place where regions adjoin.

There are eleven subspecies of the pine
grosbeak, between them covering the whole zone
worldwide. This bird is a finch but much larger
than most, being of song thrush size. As the name
suggests, the beak is large and it is adapted to
deal particularly with buds and shoots of conifers,
such as spruce, and catkins and the berries of
trees such as rowan.

Crossbills are also finches but specialize in
eating the seeds of conifers. There are several
species and subspecies. All have curved, crossed
tips to their mandibles and large, strong feet. A
bird pulls off a cone and holds it firmly between
one foot and a level branch. Then it inserts the
tip of the upper mandible below a scale and as

Above left
Perched in the top of a
conifer, a hawk owl shows
its resemblance to a true
hawk. As with other raptors,
the number of eggs laid and
young reared depend on the
available food supply; the
normal three to six may rise
to nine or more in times of
exceptional abundance.

Above right
Goshawk occur throughout
the boreal forest and also in
deciduous woodlands in
Europe, though their
numbers are much reduced
in many areas due to
persecution. Nests are
usually built high in a tree
and both sexes bring food
to the nest but only the
female actually feeds the
young.

the beak is closed the scale is forced away from the body of the cone and the seed beneath it is extracted by the bird's long tongue. When cones are ripe and open, seeds can be more easily removed. Different species have different-sized beaks and prefer different cones. For example, the heavy beak of the parrot crossbill is suited to extracting the seeds from the large, stout cones of pine, while the common crossbill's lighter beak is better adapted to the large but softer cones of spruce. The two-barred crossbill, with a rather slender beak, favours the small soft cones of the larch. The fact that the seed crops of the different trees are exploited by different species of crossbill means that competition for food is lessened and they can co-exist in the same zone.

The crossbill's breeding cycle is related to ripening of the cone crop in order that there shall be ample food for the young. Thus in the spruce forests in Europe, some birds start to breed when the cones form in August and may nest at any time thereon right through the winter months until the seeds fall in the next spring. Hatched in the winter, the young are very resistant to cold. They have been recorded in Russia hatching in an air temperature of minus 35°C. Although the temperature in the nest under the parent was comfortably warm, when the young were left they became torpid, recovering rapidly on being brooded.

Several gamebirds live in boreal forests. The capercaillie, distributed from central USSR westward to Scotland, feeds on the shoots of conifers, as does the North American spruce

Far left
The red-spotted bluethroat breeds throughout Eurasia to Alaska, favouring marshy birch and scrub habitats and usually laying six eggs. About robin-sized, bluethroats feed mainly on insects and other invertebrates as well as berries. In winter months they move south to North Africa and southern Asia.

Left
Wood sandpiper winter in freshwater marshes well to the south of their breeding ground in marshes of the boreal forest zone, birch woods in particular. Normally the nest is on the ground but sometimes the old nest of another bird is used. Soon after the eggs hatch, the hen leaves the male to rear the young alone.

grouse. Black grouse, with a range right across Europe and Asia, favour forest edges and open areas, feeding on the conifer shoots but also taking buds and catkins of deciduous trees, notably birch and other vegetable matter such as heather and berries. The hazel grouse, which has a similar distribution (though it does not extend into western Europe), favours forests with good ground cover. It eats mainly deciduous shoots and catkins, turning largely to shrub berries in autumn and to alder catkins in winter. All these species are resident throughout the winter because some food is accessible above the snow at all times.

The northern forests, like the tundra, are very rich in insects in summer. Normally overwintering as eggs, larvae or pupae, they then become harder to find. This means that only a small number of species stay in the zone and feed on insects all year round. Other insectivores migrate out of the area in winter but a third category feeds on insects in summer and the more easily obtained vegetable foods in winter.

Waxwings fall into this last group. There are three species: one occurs across Eurasia and on the other side of the north Pacific in western Canada, the cedar waxwing occurs only in North America, and there is also a Japanese species. During the breeding season waxwings take mosquitoes and other small insects, but when these become unavailable in autumn they consume berries. In most winters, as the berry stocks are exhausted, the birds move steadily south and return in the spring. Roughly every ten years, vast numbers move rapidly far south even though food is still plentiful further north.

These irruptions appear to occur when the population has reached a very high level and they are a feature of a number of boreal forest species. As many birds fail to survive or return, the process clearly seems to limit the population and though it may sometimes be triggered by overcrowding before food actually becomes short, it is probably in fact a response to actual or potential food shortage.

It is interesting to compare the waxwing's winter movement habits with those of other birds such as the redwing and fieldfare, which are both thrushes. Though they have a wider food preference than waxwings, taking not only berries in season but also insects and other invertebrates including earthworms, they are much more susceptible to cold weather, undertaking rapid movements, sometimes over long distances, at the start of a severe spell and moving back again as conditions improve. Thus, unlike most migrants, they do not return to the same region each year. Ringing of redwings has shown that a bird that visits Britain one year may be in southern Russia the next. Fieldfares may be common in one breeding area in one year and absent the following one. This flexibility allows them to take advantage of good conditions wherever they may be found.

It is interesting that the same species may have different habits in different parts of its range, and this may be a first step towards the evolution of separate species provided that there is little interbreeding between the different component populations. The great tit breeds in both coniferous and deciduous woods throughout Eurasia. When winter food becomes hard to find, great tits may irrupt from the area and in regions where winters are always difficult they may be regular migrants, while in milder zones they are resident.

A large number of warblers migrate into the boreal forest in summer. Most of these birds are small and look rather alike but there are subtle important differences in the ways in which they seek their insect food. By specializing and becoming expert in one or two techniques, each species can reduce competition with the others to the minimum, while overall the total food supply is used to the full. Thus, the North American magnolia warbler hunts for insects on the inner, lower branches of spruce trees. In the same tree there may be other birds of the same genus also hunting but in different zones—the Blackburnian warbler searches the outer, upper branches.

Titmice of several species also live on insects in the forest zone and show similar specializations. Unlike the warblers, some tits overwinter. The brown-capped chickadee of North America and the Siberian tit of Eurasia, Alaska and north-west Canada both remain

Above left
A pileated woodpecker of North America returns to the nest with a beakful of food. Like many other boreal forest birds the species also occurs in temperate woodland, its range extending south to the Caribbean. The main food is carpenter ants, which it excavates from their galleries in trees.

Below left
Waxwings periodically erupt in great numbers from their normal northern habitat and occur as far south as the Mediterranean. It is believed these movements are caused by overpopulation relative to food supply and they occur in a number of boreal species.

Above
A pair of grosbeaks at their nest. The adult male has red plumage and the female and young males are green. About song-thrush size, this is one of the biggest finches and is found throughout boreal forest in both conifer and birchwoods.

resident in conditions of extreme cold. In winter, Siberian tits move through the forest on regular circuits, usually in flocks with other tits, feeding mainly on food stored during the summer when there is a surplus, insects being killed and tucked away, together with seeds, in crevices and lichen. Though the birds will not recall the individual sites, systematic searching through the area occupied in summer is bound to reveal them.

Food storing is a technique that is used by other birds as well as titmice. It is well known amongst some crows including the boreal jays and the nutcrackers. These birds regularly hide excess food in times of plenty and thereby increase their chances of survival when it is short. The nutcrackers appear to be able to recall the locations of individual caches and use autumn-stored food to help rear their young in spring. Though they take more vegetable matter than the jays, both have a wide range of food

Left
Fieldfare at its nest in a conifer. This species is one of the few land birds to have developed colonial nesting, and predators which enter a nesting area are met by a general alarm and 'dive-bombing'. The birds do not necessarily either breed or winter in the same area each year, but move wherever conditions are favourable.

Above
The slate-coloured junco breeds in the boreal woods of Canada and southward in the Rocky Mountains and the Appalachians. In summer, seeds and insects are found by foraging in leaf litter on the ground. In winter the birds migrate south into the USA and feed in flocks in open areas where weed seeds·are numerous.

Right
The great grey shrike occurs in a wide range of habitats in northern Canada, Europe, Asia and North Africa. In the northern parts of its range it feeds mainly on small mammals, small birds and large insects. Some individuals, but not all, impale surplus prey items at a thorn bush 'larder'.

Above
Redwing, like fieldfare, take berries as an alternative to their normal insect and invertebrate prey when it is hard to find due to bad weather. The birds nest in birch and willow woods and may rear two broods of five or six young.

Right
A turkey-sized bird, the capercaillie is resident and territorial, the female laying up to eight eggs in the nest usually made on the ground, though exceptionally individuals have used old nests of other species in trees.

preferences, taking seeds and berries, insects and other invertebrates, predating nests for eggs and young and killing small mammals on the ground.

A variety of birds of prey occur in the forests. Of the diurnal species, the goshawk is distributed throughout the boreal zone as well as occurring in more southerly wooded regions. A large, powerful bird it takes a wide variety of prey including mammals up to the size of hare, through fox cubs, rabbits and squirrels to stoats, moles and mice. However, birds from a wide size range form the bulk of its prey – capercaillie and other gamebirds, ducks and waders, pigeon, thrushes, finches and titmice. There are two techniques of hunting. The bird may sit at some favoured vantage point from which it can overlook a good area, or it may cruise low through the trees and along the woodland margins. When prey is sighted, goshawks can fly very fast, quickly

picking up speed and skilfully pursuing through cover, the rounded wings and long tail making for high manoeuvrability. They are also versatile at taking prey, having long legs and large, powerful feet, and can make their captures on the ground, amongst branches or in flight. When tackling large prey on the ground they will hang on tenaciously, even grasping a fallen bough or a tussock with one foot to stop the victim dragging them along.

The hawk owl is also a diurnal raptor, and in several respects is hawk-like, with a longer tail and more pointed wings than nocturnal owls. Its hearing is also less well developed than theirs as it hunts primarily by sight. Owls provide a good demonstration of yet another way of minimizing competition–this time by hunting the same prey species but at different periods of day or night. Thus the hawk owl operates by day, the Ural owl by day and night and the long-eared owl is strictly nocturnal. Like other night-flying birds, owls' eyes are adapted to make use of whatever faint light there may be and in addition they have exceptional hearing. The long-eared owl's real ears (its name comes from the feathered tufts on top of its head) are set asymmetrically

Above
The Arctic warbler is an insectivorous species that breeds on the tundra and in boreal forest across Eurasia and Alaska. The nest is made on the ground and is usually difficult to distinguish from the surrounding vegetation. From its whole range, the species migrates to winter in south-east Asia, individuals from the European end of the range possibly travelling over 7 000 miles (11 200 kilometres) each way.

Right
A long-eared owl at its nest—usually the disused nest of some other medium-sized bird such as a crow. The long tufts which give the species its name are not true ears: these are concealed beneath the head feathers and are extremely efficient as is necessary for a strictly nocturnal hunter.

Above
A pair of two-barred
crossbills. The bills of the
newly hatched young are not
crossed, the special shape
for extracting seeds from
cones developing later. This
species feeds mostly on the
larch and its range covers
northern North America and
Asia but does not extend
into Europe.

in its head so that it can work out the position
of its prey. The direction from which a noise
comes can be calculated by judging the intensity
of sound reaching each ear. Nocturnal owls can
also tell how far away the sound source is by
working out the time lapse between the sound
reaching one ear and the other–the longer the
lapse the greater the distance. Experiments have
shown that long-eared owls can catch prey in
total darkness when they can see nothing.

The world of boreal birds is shrinking. Much
of the timber used for paper pulp and building
comes from the natural forests of the north and
enormous areas have already been destroyed.
Because timber is slow growing, few governments
have the foresight to encourage new planting
programmes until much or all of the cheap,
natural stocks have been destroyed. As a major
timber importer, Britain consumes far more than
it could grow and home-planting programmes
have now been curtailed largely because they do
not give a quick financial return.

Temperate Woodlands

The world's temperate zones contain many different regions whose character depends largely on rainfall and temperature. Where rainfall is very low or infrequent, desert conditions prevail. Where it is more reliable, grasslands develop and where rain is fairly well distributed through the year the natural climax vegetation is woodland.

Deciduous forest develops where the growing season is long and warm enough for trees to be able to afford to shed their leaves in winter, the consequent nutrient loss being quickly made up in the next summer. This means that they can develop broad leaves, most suitable for the capture of sunlight and thus for the photosynthesis that produces plant food. These leaves have to be shed in winter, when low

temperature and reduced daylight cause plant growth to slow or cease, because they permit excessive desiccation, are prone to frost damage and, by greatly increasing the structural area of the tree, make it liable to serious damage in winter storms or heavy snowfall.

The fallen leaves produce a humus-rich soil in which ground vegetation can grow rapidly in spring before the leaf canopy opens and reduces the light reaching the forest floor.

The structure of the forests varies. In different areas with different subsoils, degrees of exposure to wind or amount of rainfall, the dominant tree species will be different. Thus, in Britain, sessile oakwoods grow in the wetter west, pedunculate oaks on the heavier soils, beech favours chalk

Left
A female European sparrowhawk feeding young. Related species occur worldwide, and all hunt small woodland birds, catching them in flight. This picture shows well the bird's long legs and long tail which it uses as a rudder in flight and here for balance as it leans forward to pass food to the chicks.

Left
The red-shouldered hawk is a medium-sized North American buzzard which hunts over open forest and wooded farmland. Hunting usually takes place from a perch, the bird scanning the ground and dropping on its prey, usually rodents or other small mammals but including birds, reptiles, frogs and insects.

Below left
The woodpigeon is largely a European species which feeds on grain and vegetable matter and has become a serious agricultural pest. Like other pigeons it drinks without raising its head between beakfuls, probably so that it needs to spend the least time in a situation where it is vulnerable to predators.

Right
A male scarlet tanager feeding unfledged young. Outside the breeding season the males assume the greenish yellow plumage worn by the females. Breeding in eastern North America, particularly in oakwoods where they feed on caterpillars and other insects, scarlet tanagers move south in winter to the north-west of South America.

and limestone and willows the wetter soil. When a mature tree dies, or when an area loses its tree cover due to wind-blow or fire, different tree species may be the first colonists of the open zones. Birch often fulfils this role in oak forests, and ash with beech. Casting only light shade and being relatively short-lived these trees act as 'nurses' to the climax trees that grow at first in their shelter, to later overtake and outlive them. Even before the sub-climax trees appear, quick growing brambles, bushes and scrub will spring up in new clearings. Thus a natural forest provides a great range of different conditions,

Above left
A yellow-bellied sapsucker at its nest hole. Breeding in temperate and boreal American woodlands, the birds drill rows of small holes through the outer bark of trees and drink the sap that exudes. The species migrates south in winter as the sap flow diminishes, and turns to eating berries and fruit.

Above right
This nesting woodcock demonstrates to perfection its cryptic plumage. A wader the woodcock specializes in feeding on earthworms and the tip of its beak contains many sensitive tactile nerve endings to aid their detection. The birds are active at dusk and during the night.

with a wide species composition, structure and age of vegetation.

Originally temperate deciduous forests grew across a much larger area than they do today but for 10 000 years or more man has been clearing them to make way for his settlements and farm land, and to aid the extermination of the large carnivores that killed his livestock. Most of China's deciduous forest vanished long ages ago. Mediterranean forests still existed 2 000 years ago but much has long been destroyed, as is the forest that covered the greater part of central and western Europe. Much of the deciduous forests

of eastern USA were felled or burnt in the last 250 years. Today only a degraded relic of the former forest exists over much of these areas. Thus many of the birds of deciduous forest have either had to adapt to copse, hedgerow and garden or have disappeared from much of their former range.

In warmer, temperate zones, wherever the winters are mild, deciduous trees are replaced by evergreen species that have no need to shed their leaves to avoid frost or storm damage. However, the leaves are often not like those of deciduous species but have a thick outer skin

Left
The laughing kookaburra is a kingfisher that lives in open forest in eastern Australia and has also been introduced to south-west Australia, Tasmania and a small area of New Zealand. A wide range of animal food is taken including insects, small birds, snakes and frogs.

Above
The whip-poor-will is an American nightjar that breeds in forest clearings. Nightjars catch flying insects on the wing and hunt at night. Their small beaks open to reveal large gapes surrounded by bristle-like feathers which help in capturing the prey.

to minimize the loss of water by transpiration under the considerable heat of the summer sun. This forest type is also modified by the amount of rainfall. It is species rich and varied where the rain is copious at all seasons but where there is summer drought, the land may only support scrub.

Evergreen forest occurs or occurred in the northern hemisphere in the USA south and west of the deciduous zone, around the Mediterranean, in China and Japan. In the southern hemisphere it is found in Chile, part of Argentina, on the coast of South Africa, in the south of Australia excluding the interior region and in much of New Zealand.

Their milder winters and longer growing seasons generally make temperate forest more productive than those in the boreal zone. The individual regions hold a greater variety of bird species than do the less diverse northern forests, while there is also far more variation from region

to region. This is because, containing a wider range of tree species and ground vegetation, and varying in character from area to area, they offer more niches for animal life to occupy.

The selection by different species of distinct niches, for example a preference for different types or sizes of seed, will in turn cause even closely related species to develop different physical structures and abilities; and may modify their breeding behaviour and whole annual cycle.

All finches have beaks which are conical, served by stout muscles and attached to a strong skull. Basically most are designed to husk seeds but variation in size between species allows each to tackle different food plants. Perhaps most remarkable is the hawfinch whose powerful beak can even split cherry and olive stones, exerting a pressure equivalent to a weight of about 100 pounds (45·4 kilograms). Bird ringers know this is a bird that should only be handled in the stoutest gloves.

The goldfinch's slender and relatively long beak is best suited for small weed seeds and it can deftly extract them from thistle and dandelion heads. It is particularly interesting that the male's beak is slightly larger than the hens so he can easily extract seeds from teasel heads but she cannot. This may mean that when food is short in winter, the sexes do not compete for the same resource. To feed on such seed heads, the goldfinch must be agile, able to hop from plant to plant, to hang upside down, to pull a distant head in reach and hold it down with its foot while it feeds.

Contrast this with the chaffinch, which has a

Left
The great tit is widely distributed in the broad-leaved woodland throughout Europe and Asia in to Indonesia. It usually feeds near to or on the ground, taking insects and seeds. Depending on winter conditions, it is a resident in some areas and migratory in others.

Above
A male superb lyrebird using its strong feet to search the litter of the forest floor for invertebrates. Each male clears 'dancing grounds' in the forest where he sings and displays by hopping with tail spread and brought forward over the head. The species is polygynous.

Right
The tawny owl breeds in Europe and Asia, hunting after dark for small mammals, birds and invertebrates including earthworms. To be successful in its woodland habitat, the bird must have an intimate knowledge of its territory so the species is sedentary. Finding its own hunting area and familiarizing itself with it is vital for each young bird.

slightly stouter beak and also feeds on seeds but
only ever takes them after they have fallen to
the ground and generally does not hop but walks
or runs so that it can search faster. In the
breeding season, chaffinches feed their young
entirely on caterpillars and other insects, unlike
the goldfinch which provides mainly seeds.
Because of this, their whole breeding behaviour
differs substantially.

In summer, insects are distributed fairly evenly
in woodland, so to ensure the food supply each
male chaffinch holds a territory of 0·25 acre (1 000
square metres) or over, singing to announce his
ownership, to warn off other males and to attract
a mate. When a female appears he courts her
and if she accepts him they finally choose a nest
site within the territory. The responsibility of
maintaining the feeding territory remains a
heavy one and the male gives little help to his
mate in rearing the young. After the nesting
season and through the winter, British
chaffinches remain in or near to their territories.
At this time they are again feeding on the ground,
exploiting the vast store of dormant seeds that
lie on and in the top few inches of litter and
soil. Presumably, knowing the territory well
helps them in finding food and in evading
predators.

Goldfinches feed on seeds throughout the year
and do not hold feeding territories. In winter they

feed and roost in flocks, moving about to exploit newly ripened seed crops. As these supplies are local and sporadic in occurrence they are not easy to find. In this case, flocking is advantageous; when passing finches see others feeding, they join them and share the find so that the feeding group may grow rapidly. Flock feeding helps also in avoiding predators; many eyes share the lookout and each individual can spend more time eating and less looking round than it could if alone. Birds that feed in flocks often roost communally too and it is thought that when the roost moves off in the morning, birds which have not found much food the day before follow those that are clearly well fed and setting off purposefully to their previous day's feeding ground. Thus the feeding flock at a good site will grow quickly from day to day. When the food is exhausted the birds spread out and search widely in small groups. Again, successful birds are soon found by others.

Goldfinches find their mates in these winter flocks and pair with them then. Usually several pairs will establish very small territories close together, so that much apparently suitable nesting area is perhaps not occupied at all. Once the young are hatched, both parents seek food wherever it is to be found, often in company with other pairs, and only the immediate vicinity of the nest is defended.

Left
The azure-winged magpie is found in open woodland habitats in south-west Spain and eastern Asia. This discontinuous distribution suggests that the species has declined and thus disappeared from much of its original range. A highly social bird, it is often found in small flocks and nests semi-colonially.

If the birds have second or third broods they may move right away from the previous nest territory each time, to get close to the best ripening seed supply they can find.

Interestingly, chaffinches from northern Europe that migrate into Britain to avoid cold winters, feed in flocks like other finches. Probably this is because they do not have the detailed knowledge of a territory that resident birds have.

The range of vegetable food available to birds is not restricted merely to buds and seeds. Flowers and the nectar they contain are important for a number of temperate and tropical

Right
A cock chaffinch at its nest. One of the commonest birds of woodland throughout Europe and western Asia, it forages on the ground for a wide range of seeds but feeds its nestlings on insects. It is a migrant in the colder parts of its range but in the south it is usually resident.

Left
The Australian sitella is related to the nuthatches and like them can move freely both up and down treetrunks, whereas woodpeckers and treecreepers can only move upward. It seeks insects and spiders on the trunks and branches of trees, probing for them with its narrow beak. The nest is skilfully camouflaged with bark.

Right
The Baltimore oriole is a member of the large New World family of icterids, breeding in North and wintering in Central America. It favours mature, broad-leaved woodland especially by streams, where it feeds on insects and fruit. The bag-like nest is suspended from twigs by plant fibres and hair.

species. The Australian lorikeets feed from the flowers of eucalyptus trees, licking up the nectar with their brush-like tongues and moving nomadically from one flowering area to the next. Australia, with New Zealand, also holds most members of the large family of honeyeaters, mostly small birds which feed primarily on nectar and insects. In America, two species of woodpecker have developed the unique habit of boring holes into trees and feeding on the fresh sap that oozes out. They too have brushlike tongues to mop up the liquid.

Most woodpeckers are insect eaters. Occurring in boreal, temperate and tropical areas, all have rather similar basic structures. Their chisel beaks can cut into wood and are mounted on a skull which is specially constructed to withstand the stresses of hammering, while the whole unit is powered by strong musculature. To provide a firm working platform, the woodpecker hangs on the trunk of the tree, sharp claws gripping into the bark, usually with two claws forward and two back, and leans back on its tail, the centre shafts of the feathers being strong enough to support it. From this position it can hammer effectively and, once the bird has broken through into the galleries of boring insects, it can probe with its tongue along them for up to four times the length of its beak – that is, up to 5 inches (13 centimetres) or more depending on the species. Backward-pointing barbs at its tip harpoon the prey and pull it out as the tongue is withdrawn.

The woodpecker's extensible tongue is mounted on a structure of elastic tissue and flexible bone which passes in two strips around the back of the bird's skull and over the top of the head, and is anchored at the right nostril. This produces a spring-like device, muscular contraction forcing the tongue out and relaxation allowing it to withdraw very quickly.

Some woodpeckers do not seek food only in trees. The European green woodpecker and the North American flicker both feed extensively on ants. The green woodpecker secretes a sticky saliva so that when it extends its tongue on to an ants' nest, the ants that attack it get stuck and can be easily eaten. In winter, when the ants retreat underground, green woodpeckers will break down the nests to reach them.

Some species take acorns, beech mast, hazelnuts, cones and similar vegetable foods.

Great spotted woodpeckers and others excavate or use natural clefts in which to wedge such items so that they can conveniently hammer them open. Nestlings of other birds are also eaten occasionally.

Woodpeckers use their wood-boring skill for a number of purposes – seeking food, excavating a nest hole and 'drumming'. In drumming, the bird beats rapidly on a carefully selected branch which will resonate, so that the sound is amplified and will carry for a long distance. It occurs in the spring and is clearly a substitute for song, allowing male and female birds to make contact and indicating territorial possession. In some species both sexes drum and, although the noise is made mechanically, all individuals of the same species produce sounds of practically the same pitch and rate, while each different species creates a different effect, just as other birds have differing songs. However, not all species rely on drumming; the green woodpecker has a carrying voice, its ringing *yaffle* announcing its presence.

Quite a number of birds nest in holes, seeking the natural cavities in old wood or excavating in rotting timber. Woodpeckers are all hole nesters and may attack healthy trees, boring first in and then down to produce a nest site which

Below left
The red-eyed vireo is a secretive but very common bird of American deciduous woodland. Foraging in the canopy for insects—and in autumn for fruit—the species nests much lower down in understorey trees. It migrates to northern South America in the winter months.

Right
The American robin breeds through the boreal and temperate woods, farmland and gardens of America. There is a southward movement within the subcontinent in winter. With the size and build of a European blackbird, this species has similar habits and food preferences—invertebrates and fruit.

is not readily accessible to predators. The red-cockaded woodpecker, which nests in pine trees in temperate woods in south-east USA, surrounds the bark for several feet up and down the trunk at its nest hole with small holes that penetrate the bark and allow resin to exude. Thus the whole zone becomes one sticky mass which is thought to deter the approach of predators.

Other, unrelated birds have developed the ability to feed on vertical trunks. Treecreepers climb nimbly, looking almost like mice, using the stiffened shafts of their tail feathers for support and probing with their slender awl-shaped beaks in the fine crevices of bark and lichen for insects and spiders. Nuthatches are the only group of birds able to move equally well either up or down a trunk. Instead of climbing vertically, hanging by their toes and propped on their tails, they move with the feet positioned one above the other and pulling against each other. Their tails are not stiffened because they do not use them for support.

Temperate woodlands hold a far wider variety of insectivorous birds than boreal forests. There is a similar stratification with different species feeding at different levels. It is particularly interesting that in temperate forest the bulk of birds find their food not in the trees but in the shrubs and plants that grow beneath them. This is because the tree layer is relatively poorer in foods—particularly fruits and insects—than the undergrowth. It is very noticeable that pure

stands of beech, for instance, which cast deep shade in which few flowers can grow, except in early spring before the trees come into leaf, support much fewer birds than the more open woods. One result of this is that it tends to be in the areas where the tree canopy is broken in clearings and at the edges of woods that most birds find food. Well-grown hedges may be much like this 'edge' habitat and many woodland birds species have been able to survive where hedges remain in the landscape.

A wide variety of predatory birds live in woodland, feeding on the surplus production of other species. It is clear that predators do not normally control the numbers of their prey but rather the reverse. When the prey population is high predators live well but when prey numbers start to fall, perhaps because their food becomes short, predators too have more trouble in finding enough to eat and they decline in their turn. Some, such as jays, take animals—mainly eggs and chicks—only as a secondary food, eating large numbers of insects and depending primarily on acorns in autumn and winter.

Man's activities, particularly in agriculture, have enormously modified the temperate forest habitat. Much of it has been destroyed so the many birds that depend on it have declined. However, the open land that has replaced it, particularly where it contains hedges which can reproduce the conditions of wood edge, providing song posts, nest sites, roosting places and food, has favoured some woodland species. Until herbicides reduced the weed populations and thus the food supply, goldfinches found farmland a most favourable feeding habitat. Sometimes man's activities have produced conditions which particularly favour the natural preference of one species and this can create genuine problems because the bird is likely to rapidly increase in numbers and become a pest. Woodpigeons have profited immensely from man's consideration in growing great quantities of green stuff in the winter when natural foods are short. Bullfinches are a major pest in fruit-growing areas where the flower buds provide a lavish alternative food supply in years when ash seeds are scarce. Cypriot grapes provide migrant blackcaps and other warblers with an energy rich food on autumn migration.

Other conflicts have arisen. Australian nectar-feeding lorikeets compete with apiarists' bees for the raw material of honey. Game preservation retains woodlands but often destroys the raptors and other predators that are a proper part of natural bird communities. In many countries, sport shooting and trapping for food or caging is popular.

Even where woodlands remain, their natural character may be severely modified. Management for timber reduces greatly the amount of old or dead wood, with its nesting cavities and rich invertebrate populations that provide food. The growing of alien species such as conifers produces new habits to which the indiginous avifauna may not be adapted.

Most temperate forest is in 'developed' nations. If world trends continue, at least European countries will seek greater agricultural self-sufficiency and this is likely to lead to more loss of the remaining woodlands and of the hedge and other marginal habitats in which many woodland birds still survive today.

Above
The white-naped honeyeater feeds on insects and the nectar of eucalyptus trees in eastern Australia, moving nomadically to new areas as the trees come into flower. The nest is a cup made of bark, fibre and cobwebs, lined with grass and suspended from a branch. Two or three young are produced.

Tropical Forests

Tropical forests grow where temperatures remain high right through the year and there is generous rainfall. The hot, wet climate supports a rich growth of trees and other vegetation. In a band extending to about 3° north and south of the equator, heavy rainfall occurs fairly equally in every month averaging about 160 inches (406 centimetres) for the year (compared with about 25 inches (64 centimetres) a year in London). This is the true rainforest belt. Further away, up to about 10° north and south, the rain is heaviest in June and December but may still amount to 10 inches (25 centimetres) a month in the driest periods. These two zones together contain the great Amazonian forest of South America, the forests of west and central Africa and those of Indonesia.

Further north and south still, the rains are concentrated in one short season—the monsoon—and there may be a well-marked dry season in which trees shed their leaves. The forests of India, south-east Asia and north Australia lie in this zone.

The structure of tropical forests is the same in all continents. Trees are dominant, not merely in size but sometimes also in numbers and variety. They usually form a continuous closed canopy whose roof is about 120 feet (37 metres) above

Left
The black-mantled goshawk occurs in forest up to high altitudes in the mainland of New Guinea, feeding not only on birds but also frogs and insects. Birds of prey are less common in tropical forest than in grassland because of the greater difficulty in finding and capturing prey there.

Right
Feeding on the ground, where it rakes through the litter to find insects and also eats fruit, the Australian brush turkey builds a nest mound of wet vegetation in which the eggs are incubated by the heat of its decomposition. The bird's bare head is regularly thrust into the mound to assess its warmth and vegetation is removed or replaced as required to keep the temperature right.

Below right
The white-winged trumpeter is a weak flyer and lives on the ground in the Amazonian rainforest. Often forming large flocks, the birds may follow beneath foraging troops of monkeys picking up the fruits and other foods that they drop. For safety they roost in the lower branches of trees.

the forest floor. Here and there still taller giants stand out over the otherwise unbroken and usually sunlit sea of leaves. Beneath it the under-storey is in shade through which the trunks of the large trees rise unbranching for 50–80 feet (15–24 metres). Where the canopy is not too dense, or in open spaces where a tree has fallen, the ground may support a herb and shrub layer as well as an intermediate layer composed of young trees, tree-ferns and palms. Epiphytes, taking nutrients from the rain and humid air, grow on the trees, and lianas form a tangle of rope-like stems.

Within the broad category of tropical forest there are many variations. On its fringes, as the forest approaches the savanna, its height, diversity and density may be reduced. In mountain regions, increasing altitude affects the climate and thus the vegetation just as does increase in latitude, so that tropical growth is replaced by woodland of temperate type, and so on. Low lying areas may be flooded during the rains and support waterbird populations some of which, like the tiger-herons, carry out regular movements from area to area, feeding in places newly exposed by the receding flood waters.

The high temperatures and rainfall create an extended or continuous growing season so that

Left
The red-legged falconet is the smallest bird of prey, being no bigger than a starling, and occurs in south-east Asia. Frequenting forest clearings where it can command a clear field of view, it captures large insects and occasionally small birds in flight.

Below
The nest of the common palm swift is stuck with saliva to the hanging frond of a palm tree and the eggs are similarly stuck into the nest. The incubating bird must cling on vertically. With their very long, slender wings, swifts spend most of their lives on the wing— preening, mating and sleeping in flight.

plants may be producing food for animal life all year round. This primary production by the plants is exploited by great numbers of invertebrates. Ants and termites are particularly successful and, if the humidity is high, leeches and flatworms can live out of water in the terrestrial environment. There are great numbers of different plant species in these forests–perhaps as many as 3000 different kinds of tree in Indonesia alone–so that their characteristic is variety. This applies equally to birds. As a result of the great number of specialized niches available, many species have evolved in the tropical forests. A count in Ecuador found 200 different species occurring in one 12 acre (48000 square metre) area, and this makes a vivid comparison with the British situation where 200

Left
The powerful notched beak of the palm cockatoo can break open the hard shells of nuts, and as the male has a larger upper mandible, it can probably eat nuts the female could not handle, thus reducing competition between the sexes. The species is found in New Guinea and north Queensland.

Right
The jungle fowl of south-east Asia is probably the species from which all domestic poultry descend. The birds seek insects, seeds, berries and other plant foods such as shoots on the ground, having strong legs and feet with stout claws to help them scratch through the litter and soil.

is a respectable species total for a full year's birdwatching taking in all habitats, not just woodlands. As a consequence, no one species can be particularly common in one place. There is a striking contrast with boreal forests where there are much fewer species and one alone may dominate vast areas.

The separate forest avifaunas of the different continents are very largely distinct–again in marked contrast to boreal forest where many species occur right round the world–but because the forest structure and opportunities are similar, birds with quite different origins have evolved similar adaptations and this convergence can be found in almost all groups.

There are few fires or wind storms in tropical forests and the primary forest structure is one of great stability and antiquity which has provided unchanging conditions for many millions of years. As a result some of the bird species which live in it are themselves of ancient lineage–for example, the Congo peacock of Africa and the hoatzin and trogons of South America.

Where the forest has been destroyed for primitive agriculture the soils, which are notoriously poor, are quickly exhausted so that the cultivators must move on to another patch and secondary forest springs up with a less enclosed canopy and even higher production of plant and insect food. Many virgin forest species cannot adapt to this habitat but it nonetheless supports a large and varied avifauna. It may take hundreds of years for primary forests to redevelop.

Because temperature and food supply are always adequate, tropical forest birds are largely sedentary, having no need to migrate out to avoid temporarily unfavourable conditions. Also, because there is no seasonal superabundance of food, the habitat supports few incoming migrants.

Despite the heat and humidity so apparent to the unacclimatized human visitor, birds appear to find no special problems in regulating their body temperatures. By confining most of their activity to the hours just after dawn and before nightfall they avoid needless exertion in the hotter part of the day and, of course, this means that they feed in two periods about twelve hours apart as do birds in many other regions. Their other main concession to the climate is to generally stay in the shade, thus avoiding the direct heat of the sun, and to bathe frequently.

The forest can be considered as a number of fairly well-defined zones placed one on top of the other in layers, each with a distinct bird community that may not venture even into the adjoining strata. At ground level are birds, often with weak powers of flight, that glean food from the litter of leaves on the forest floor. Many have powerful legs and feet for their rummaging and will run rather than fly when disturbed. Some pheasants may have a top sprint speed of over 20 miles per hour (32 kilometres per hour). The cassowaries of New Guinea and Queensland have abandoned the ability to fly in exchange for large size (they stand up to 5 foot (1·5 metres) high and weigh about 1 hundredweight (51 kilograms) and great running ability.

Off the ground in the bushes and small trees, the birds are mainly insectivorous and primarily arboreal, though some do descend to the ground. There are species adapted to catch insects in flight, to search for them amongst foliage and on or in the bark of trees. The 'tree' kingfishers, such as the banded kingfisher of south-east Asia, are notable as they have abandoned the aquatic habitat of their more familiar relatives. The banded kingfisher hunts from a perch overlooking an open space in the forest and pounces on lizards, frogs, small insects or other small prey

Left
The hyacinthine macaw of the Amazonian forest is about 3 feet (0·9 metre) long from head to tail tip and flies powerfully, moving through the top of the forest canopy and probably feeding on nuts, fruit and berries. They particularly favour the dense forest belts that grow along the banks of rivers.

Above right
Cock-of-the-rock males gather in groups where each bird displays on his own perch and a cleared patch on the ground flying from one to the other and fanning the crest forward to conceal the bill. After the female has chosen a mate, she incubates and rears the young alone.

Below right
Great hill barbet at a nest hole. Barbets are a widely distributed family in Asia, Africa and South America. Allied to woodpeckers, they have the same 'two forward, two back' toe arrangement, strongly shafted tails for support and powerful beaks which they use to excavate the nest. Most species feed primarily on fruit.

Above
The blue-crowned motmot is
a bird of open tropical forest
in America. Feeding mainly
on insects, motmots are one
of a number of species that
will follow columns of army
ants in order to feed on the
insects that they flush from
concealment. The birds nest
in tunnels that they excavate
in moist soil and which may
be up to 7 feet (2·1 metres)
long and take eight months
or more to complete.

Right
The great horned owl is a
nocturnal species found in
all types of wooded country
throughout America, being
least common in tropical
lowlands. Standing 1·5–2
feet (0·4–0·6 metre) tall, it
takes live prey up to rabbit
size. Outside the breeding
season the male and female
probably live separately in
individual hunting territories.

on the ground or in the trees as well as taking
insects in flight and excavating rotten wood for
beetle larvae.

Many insectivorous species join in large mixed
flocks on a scale far greater than that seen,
particularly involving tits, in temperate and boreal forest. Several hundred individuals–flycatchers, thrushes, sunbirds, warblers, woodpeckers, parakeets and others–may come from
their overnight roosts to form a single party and
forage together. The whole flock moves through
the forest over regularly used routes, its individual components feeding in their individual
ways–some on or near the ground, others in the
higher layers or flycatching on the wing. All
benefit from the association primarily because the
disturbance caused by the flock stirs up all the
insects in the area and makes many more of them
visible to the individual birds which, continuing
to feed in their different ways, are not competing
with each other.

These flocks are socially organized and appear
to have a dominant leader, while the members
understand the meaning of at least some of the
calls of other species, particularly those which
signal the assembly of the flock and those giving
alarm. The ability to recognize calls denoting the

Above
With a distribution similar to that of the palm cockatoo, the eclectus parrot is thought to feed more on fruit and nectar than on very hard nuts. The female is quite unlike the male which is largely green; her head and neck are red, back and underparts blue and wings and tail are maroon.

Left
The toco toucan is one of several species found in tropical forests of America where they feed on fruit and insects. The large beak is quite light as its internal construction is cellular. It is not known whether the beak was developed primarily for display or to aid feeding.

approach of a predator is valuable not only to birds in flocks but to any individual and in fact many different species have similar alarm calls.

Flocks will also travel with columns of army ants in order to feed on the other insects and small creatures that the ants disturb. Monkeys also cause exploitable disturbance as they travel through the canopy; hornbills and drongos will follow them.

Apart from insects, fruit and nectar are the main food sources for birds in these forests. Seeds are a relatively small resource and leaves are eaten by only one species–the primitive hoatzin. Fruit grows mainly in the canopy where parrots, toucans, fruit pigeons and many other species feed on it, moving around from one part of the forest to another as different crops ripen. This soft food requires no special beak adaptation and birds with very differing bills feed on it. However it has been suggested that the large beaks of toucans and hornbills enable these large birds

to reach out for fruit growing on branches too slender to take their weight. Fruit pigeons, such as the green imperial pigeon of south-east Asia, have very wide gapes that can accommodate fruit such as nutmegs up to 2 inches (5 centimetres) across, while their short, wide intestines will easily pass the large stones. With flocking habits and communal roosts, maximum numbers can exploit each new-ripened food supply as it is found, in the same way that finches do with seed supplies in temperate areas.

Nectar is an important and energy-rich food available from the flowers of many trees and eaten by hummingbirds, sunbirds, flowerpeckers and others. There are over 300 species of hummingbird, all in America and most of them tropical. Not only capable of hovering, they are the only birds which can really fly backwards, their wings beating far faster than other birds and in the smaller species reaching eighty beats a second. There is great variety of beak shapes

Left
Found only in the mountain forest of Central America, the resplendent quetzal nests in holes in rotting trees. The adults feed largely on fruit which they pluck in flight, but the young are fed initially on insects, invertebrates and small reptiles. When incubating, the male's long tail is left sticking out of the nest hole.

Above
The violet whistling thrush lives alongside streams in forested mountain areas of south-east Asia. Its white-speckled blue plumage provides effective camouflage in the shadows beside the water where the birds search for insects and invertebrates, also flying up into the surrounding trees to take berries and seeds.

Left
Gould's manakin, like the cock-of-the-rock, is a species in which the males display communally and the females rear the young unaided. Such behaviour is only possible in species with an abundant food supply where one adult can cope adequately. It allows the most vigorous and successful males to mate with several females and thus helps natural selection.

Below left
The grey-necked bald crow lives in colonies in large open caverns and is confined to the western part of the Congo rainforest. The selected sites are all used by bat colonies and the birds feed at least in part on the cockroaches and other insects which eat the bats' droppings. Building nests within the caves, they leave them only to venture into the dense ground vegetation that grows at their entrances.

designed to penetrate different shapes and sizes of flowers, the sword-billed hummer having a beak as long as its body. These birds have tubular tongues with which to suck their food. With a very high metabolic rate, the hummer's energy requirement is great. For comparison, the food value of its daily nectar intake may equal a daily consumption by a human of 3 hundredweight (152 kilograms) of boiled potatoes. The similarity between hummers and certain large moths is yet another example of the way in which totally unrelated creatures that adopt similar modes of living develop similar physical characteristics.

Birds of prey are not particularly numerous in tropical forests because it is a difficult habitat in which to hunt, other birds, reptiles and mammals having so many hiding places and opportunities for evasion. A few large species such as the harpy take monkeys, squirrels and occasional birds. The bat hawk is particularly interesting. A kite, with the build of a very large falcon, it takes small bats and hirundines mainly in the brief dusk, seizing them in its feet and swallowing them whole while in flight in order to make maximum use of its brief daily hunting period. One bird of prey, the palm nut vulture, has largely abandoned a carnivorous habit to feed almost exclusively on the fruit of the oil palm and the raffia palm.

Many tropical birds are renowned for their brilliant colours but these are mostly confined

Above
Like many American tropical forest birds, the bare-throated bellbird has been little studied in the wild. It and the white bellbird are among the very few species of land bird which have unmarked white plumage.

to species that live in the bright canopy layers. The birds of the undergrowth are usually far duller and often cryptically patterned. It seems most likely that the bright plumages are also cryptic amongst the moving sunlit leaves.

There is no major seasonal temperature change in the tropics and breeding is usually synchronized with the wetter season when forest productivity is at its highest. Clutch size in the great majority of tropical birds is only two eggs, and this compares oddly with the much larger broods of birds of colder areas where the natural food productivity is lower. It seems likely that, living sedentary lives in stable and rich habitats, tropical birds are longer lived than many other species and thus need to produce fewer young to maintain their numbers.

As the individuals of any one species are likely to be widely scattered, song is particularly important in attracting a mate and perhaps less so for territorial defence than is the case in temperate habitats where many individuals of the same species may occur in a small area. Many birds have very carrying songs and it has been reported that the ringing calls of the male bellbird can be heard at a range of 3 miles (5 kilometres), though this seems unlikely except under abnormal conditions. Members of a number of unrelated families have developed duetting in which both sexes sing together or successively to produce a single song and this joint involvement must help to reinforce the pair bond.

In most birds, however, it is display which brings the female to a condition in which she is ready to mate. In some cases the bird may be incapable of producing eggs until the stimulus of courtship triggers the necessary physiological mechanisms.

There are many variations on the basic theme. For instance, many male and female hummingbirds hold separate territories, the male's being larger and extending into those of several females. Initially he attracts a female by an elaborate flying display, which she then joins and the pair perform an aerial dance which becomes increasingly skilful as the two birds

Above
Found in Burma, Malay Peninsula and Indonesia, the green broadbill is a fruit-eating species. It constructs a gourd-shaped nest of dead leaves and fibres, suspended from a branch usually over a stream where it looks like debris left hanging by receding flood waters.

Right
The wedge-billed woodcreeper lives in tropical American forests where it seeks insects like a treecreeper, climbing round and up a trunk and then flying down to start afresh at the base of another. The different sizes and shapes of the beaks of related species enable each to feed best on a different size of insect.

Above
Outside the breeding season, the magpie tanager often occurs in the large mixed flocks of birds which move through the tropical American forest. This species feeds on ripe fruit but other members of the flock will take other foods of many types.

Above right
Widely distributed in African tropical forest and savanna, the green pigeon lays a single egg in the forest habitat and usually two elsewhere. This suggests that because of the stable fruit supply in the forest, plus the relatively low density of predatory birds, individuals that breed there live longer and thus need to produce fewer young to maintain their numbers.

Right
A social species, the American yellow-rumped cacique feeds in flocks on fruit, berries and insects and breeds colonially. The females build closely spaced, untidy, hanging nests of fibres and carry out incubation and rearing of the young without help from the males. The yellow-rumped cacique prefers forest fringes and open country while the red-rumped cacique penetrates deeper into the tropical forest.

become familiar. At last the pair bond reaches the point where mating is possible. After this the female takes sole responsibility for building the nest and rearing the young. It is likely that, as nectar is a readily available food supply, the attentions of only one parent are enough to raise the young and a number of other families of fruit or nectar eaters manage similarly. In this instance it is the male who is successful in holding territory, that breeds and passes on his good genetic traits.

In certain manakin species of South America the males stage communal displays, with up to seventy birds each holding a few square feet of individual territory and all displaying excitedly whenever a female comes on the scene. The display grounds are traditional so that contact between the sexes is facilitated and in 'arena' species it is one or a few dominant males of superior energy and aggressive drive which mate with all the visiting females.

To enhance their displays, many male tropical

Above
Woodland kingfishers are found in African forest and savanna where they favour edge habitats along streams and beside cultivation. Related species occur in densely wooded or more open habitats. Eating mostly insects and lizards, they fly down from an observation perch to seize prey on the ground.

Right
Found only in the Philippine Islands, the monkey-eating eagle is now very rare. Having broad, rounded wings and a long tail, it resembles a massive hawk and has a flapping flight rather than soaring. It catches monkeys, lemurs, squirrels and large birds in the treetops.

birds have developed plumages which are amazingly structured as well as or instead of being brilliantly coloured. Few can be more bizarre than the birds of paradise. The male blue bird of paradise actually hangs upside down in order to display the long and finely filamented feathers of his flanks.

The bowerbirds are related to the birds of paradise and males build elaborate structures on which to base courtship. Different species construct different styles of bower—some consisting of heaped up cones of twigs joined by a display perch, while others make parallel walls of twigs nearly a foot (30 centimetres) high and about 4 inches (10 centimetres) apart. Bowers are often ornamented with collected objects, some species favouring lichens or flowers, others selecting items of one colour only or 'painting' the walls with a pulp of vegetable matter and saliva.

Even saliva itself has been adapted by some birds to allow them to occupy niches that other birds cannot. Swifts use secretions from the salivary glands in the construction of their nests. Designed for high speed and manoeuvrability, taking insects from the air above the forest canopy, swifts have such long, slim wings that they cannot fly through cover and can hardly take off from the ground at all. They, therefore,

need to construct their nests where they have a clear approach and departure path. The common palm swift of Africa and south-east Asia uses saliva to glue its nest to the underside of a hanging palm leaf and also glues the eggs into the nest. The bird itself has to cling on vertically to incubate, as do the nestlings especially in winds when the nest may even be blown upside down.

Perhaps the most famed of nests is that produced by the edible nest swiftlet of south-east Asia, a saucer-shaped construction built up entirely of salivary secretions and stuck by one edge to the wall of a cave. Nesting in what may be total darkness, these swiftlets and the oilbirds of South America use echo-location to find their way and avoid bumping into the walls. Emitting successive 'clicks' at a rate of about 250 per second, they can judge precisely from the time taken by an echo to return from the solid surface how far away it is. Where many individuals are flying in one place each must emit a slightly different sound so that it can distinguish its own echoes from those of other birds.

By coincidence, both these species have long been exploited by man. Today the edible nest trade is regulated so that the nests are not supposed to be collected until after the young have fledged. Oilbirds, whose fruit-fed chicks become hugely fat and were once collected to provide oil for cooking and lamps, are today protected in at least part of their range.

The conservation of tropical birds depends, as elsewhere, on the conservation of their habitats. Regulation of the uncontrolled and unquestionably callous trade in tropical birds, largely for Britain, western Europe, Japan and the USA, is certainly needed on humane grounds and to protect a number of very rare species. However, the real problem is that much of the greater part of these forests grows in poor countries which cannot support their rapidly growing populations without exporting their timber resources and then, all too frequently, using the cleared land for agriculture which is often doomed to failure after a few years because the soils are so poor. War too has become a significant factor, as the massive devastation of Vietnam's forests bears witness. But since western nations were intimately involved in the destruction of Vietnam and tactfully close their eyes to the extermination of the indigenous and little known peoples of the Amazon basin, what price the birds?

Left
The fairy bluebird of south-east Asia is found mostly in small flocks, often with other species, moving through the forest canopy where it feeds on figs and other fruit. The nest is a shallow, rather fragile-looking cup of twigs and moss placed in a bush or tree and both adults cooperate in rearing the two young.

Above right
Preferring semi-open habitats, the glittering-bellied emerald is found on the edges of forest clearings and tropical savanna of South America. It feeds partly on insects and nectar, to reach which it will pierce through the throat of flowers with trumpets too long for its beak to reach down.

Below right
Feeding mostly on termites and other insects, barred antshrikes inhabit the undergrowth of secondary forest and clearings in tropical America. Each pair is territorial and, as in many other tropical birds, breeding may occur in almost any month because conditions vary so little throughout the year.

Savanna and Steppe

Used to living in Britain, where the landscape changes almost from hour to hour as you drive across the country, it is difficult to grasp the sheer extent of some habitats. The Eurasian steppes start in Hungary and stretch on eastward across the southern USSR to China for 2 000 miles (3 200 kilometres). The prairies of North America, lying in the western half of the continent, extends south from the boreal forest almost to Mexico. In South America the Argentinian pampas covers 200 000 square miles (500 000 square kilometres) between the Atlantic and the Andes. The veld of southern Africa, the grassland belt of eastern Australia and the downs of eastern New Zealand are less extensive but still extremely large. These are the world's temperate grasslands, with hot summers, cold winters and little rainfall at any season so that only a few scattered, drought-resistant trees may grow.

Grasslands also occur on an equally large scale in tropical and subtropical zones where, though the temperature remains fairly high all year, rain falls only in summer. Forming a transitional zone between high forest at one extreme and desert at the other, tropical savannas may have a fairly extensive cover of palm trees and other vegetation or, in the dryer areas, carry only small and scattered thorn bushes. In Asia they occur in much of central and northern India as well as parts of the Far East. In America, they lie both north and south of the Amazon rainforest, forming the llanos in Venezuela and much larger campos of Brazil, which on its southern boundary reaches the pampas.

The African savannas cover about four million square miles (ten million square kilometres) in a belt running across the continent south of the Sahara, down the east side and back to the west

Below left
Though it favours open country, the black-shouldered kite occurs in many habitats and every continent, being commonest in Africa, south-east Asia and Australia. Hovering, soaring or gliding, it searches for large insects, small animals and birds and will occur in flocks where food is plentiful—for instance, following locust swarms.

Right
The stonechat is widely distributed, being largely sedentary in the African grasslands but migrating south from much of its European, Siberian and mid-Asian breeding habitat into tropical areas when its insect food becomes seasonally unobtainable. Here a cock bird adopts a typical posture on an exposed perch.

coast to the south of the Congo rainforest. Savanna also covers western Madagascar and much of northern Australia.

The amount of food available to birds in these areas varies considerably from season to season. When the rains come–in spring in the steppe and during summer on the savanna–there is a flush of new plant growth and a consequent increase in food. If the rains are very heavy, as they are in some areas, resident birds may actually be forced to move away at this time but, once they cease, great numbers return. Because of the nature of the vegetation, the food it provides for birds is mainly in the form of grass seeds, there being few soft fruits or nectar-bearing flowers,

Above
The Egyptian vulture is one of a minute number of birds known to use tools. The bird will pick up a stone, carry it to an untended ostrich nest and drop it repeatedly on an egg until the shell breaks and the contents can be eaten.

Right
Cockatiels are found in most open habitats in Australia moving nomadically to wherever they can find food—mostly seeds but also insects—and water. Breeding after the rains, the nest is made in a hole in a tree.

Far right
A long-crested helmet shrike incubates on its nest of fine grasses and cobwebs, which are also used to anchor it in position. This species is widely distributed in the African wooded savannas and commonly occurs in small flocks, feeding in trees and on the ground, largely on insects.

Above
The crowned crane is found throughout temperate and savanna grasslands of Africa though it does not usually disperse far from rivers or marshes. In the breeding season, pairs nest separately but otherwise the species is gregarious, flocks concentrating where seeds, insects or reptiles are plentiful.

Left
Ostriches would suffer from an overheating problem, because of their great size, if they did not have large areas of exposed skin. Here a male displays his wing and tail plumes to a female. Ostriches have only two toes and their feet resemble the hoofs of herbivorous mammals.

Top
With their powerful beaks, two lappet-faced vultures prepare to tackle a hyena carcass. The waiting white-backed vultures and marabou storks make way for them. Vultures have a feeding hierarchy, the large, heavy beaked species being best suited to breaking into a carcass while bare-necked species can thrust their heads inside to feed.

Above
As it alights, the African dark chanting goshawk displays the round wings, long tail and long legs that typify woodland hunters. However, this species has adapted to open savanna and mostly takes lizards and insects on the ground.

but this does not prevent insects from also being abundant. Large areas of low lying ground may be flooded creating temporary wetland habitats.

Then the drying out begins. The marshes shrink and disappear so that the water birds leave. Grasses dry in the wind and trees shed their leaves. Insects become scarce and even seeds are hard to find. In the steppes the winter is cold. There is, therefore, a great cyclic movement of birds, not only within these areas but also into and out of them. In spring, the temperate steppes receive many migrants from wintering grounds in the other hemisphere. In summer, the tropical savannas hold not only breeding birds but also great numbers of 'wintering' birds which have moved away from breeding grounds across the equator that are now too cold and short of food to hold them. In this respect, the tropical savannas contrast with the tropical forest. In the latter, food is equally abundant all year round and the bird population is resident with nothing to spare for seasonal immigrants.

The lack of trees over huge areas has major effects on the birds' breeding habits. Nests on the ground are particularly vulnerable, not only to predators but to the feet of grazing herbivores, and in many species the young leave their nests as soon as they are hatched in order to reduce the risk of total loss. Birds of prey, which mostly favour tree or cliff sites for breeding, are less common in areas far from such sites but some, such as the steppe eagle, will build on the ground.

Raptors usually advertise their ownership of a territory by display flights but most small birds rely on song. Deprived of perches, larks and others ascend almost vertically into the sky and 'parachute' down as they sing. They build open nests on the ground but nonetheless seek maximum concealment and shelter from sun or rain. Burrows are attractive for this reason. The American burrowing owl digs its own or lives in holes made by prairie dogs, skunks or badgers. Bee-eaters and grassland kingfishers excavate tunnel sites in banks. The red ovenbird of South America builds its own weatherproof nest–a rough sphere of mud, grass and hair that drys like cement. The whole structure is about 10 inches (25 centimetres) across with walls 1·5 inches (4 centimetres) thick and an entrance hole at the side. It is placed on any available support–tree branch, telegraph pole or, where no alternative exists, on the ground.

Other birds have devised weather and predator-resistant nests in quite different ways.

Left
Bateleur eagle incubating its single egg. With very broad wings and short tail bateleurs, like albatrosses, can glide far on even slight winds or thermals, thus economizing on energy, while searching for food on the ground. They take a wide range from small antelopes to large insects, snakes and carrion.

Above
Wintering in India, black-headed buntings move west to breed in dry temperate grasslands and rocky hills between Italy and Iran. In a partial post-breeding moult, the males lose their black head colour and in a full winter moult become brownish, gradually donning breeding plumage as the brown edges of the feathers wear away to expose the bright colours beneath. This saves the energy that would be needed for another full moult in spring.

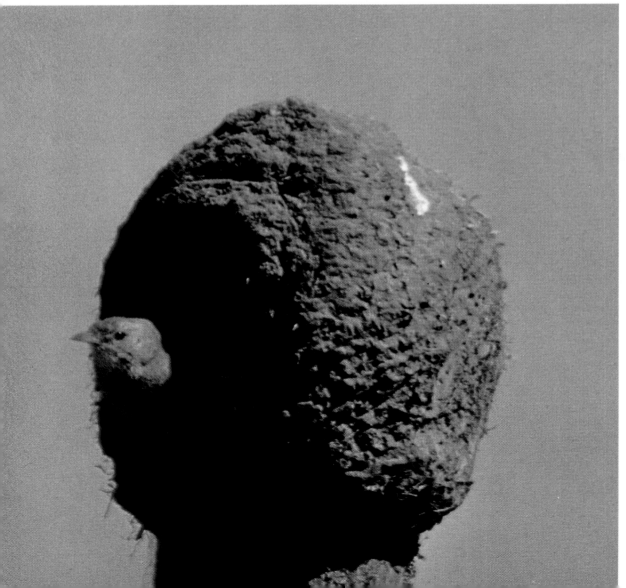

Above left
The American prairie chicken is a large gamebird once common but now extinct in many areas and rare in others due to heavy hunting pressures and loss of its habitat to agriculture. The males display at communal leks, inflating air sacs on their necks and making a booming call.

Left
A rufous ovenbird peers out from its nest. Constructed of mud and drying to the consistency of cement, the nest provides a weatherproof and predator resistant solution to the shortage of sites in the open grasslands of South America.

Above right
Throughout the centre of Australia, but usually not far from water, budgerigars are found feeding on seeds, which they take from the plants and pick up on the ground. Like many other species in the area, they appear to be nomadic and huge numbers may occur temporarily where conditions favour them.

The long-tailed tailor bird of India and south-east Asia sews together the edge of a single large leaf or brings two leaves together edge to edge to form a funnel in which the nest is placed. Suspended from a thin bough, it is not easily accessible to predators and as the leaves remain green, they help to conceal it.

The African weaver birds construct their nests entirely from strips of grass which are skilfully woven and knotted in one of a number of characteristic designs. Most have a thick, rounded roof to shed water and a tunnel entrance hanging downwards to defeat predators. The nest may be suspended from the tip of a twig growing out over water for further protection. Some species even strip off nearby leaves, perhaps so that no snake may approach undetected.

Choice sites such as those over water are in relatively short supply and this has affected the breeding ecology of different species in different ways. Many are colonial and nest close together. Some build a massive communal structure divided into nesting compartments and this economizes on the individual effort and time needed during construction. As a nesting colony is more likely to be noticed by predators, some weaver species only next colonially in safe sites over water,

becoming solitary nesters relying on concealment if they have to breed in reedbeds.

In many seed-eating weavers, the habit of colonial nesting in secure sites seems to be associated with polygyny – one male having several mates. It is possible that this arose initially because there were less suitable nest territories than could accommodate all the birds. Thus, there were numbers of females unmated, and males which had obtained territories and then accepted extra females into them would increase their own productivity. This is desirable in terms of natural selection because the successful males would be passing on their genes to more offspring than would otherwise be the case. This system leaves many unmated males, so competition for territories would be intensified, leading to progressive elaboration of plumage and display. In turn, the consequence of this would be to make breeding males especially obvious and vulnerable to predators; the fact that outside the breeding season they moult into a drabber plumage resembling the females' would seem to support this theory. The result of predation would be a reduction in their numbers so that females would be somewhat more numerous than males. At the same time young

males, if they tried to breed, while equally or more liable to predation would be less successful at holding territories than the more experienced two-year-old birds. As a result, one-year-old males do not attempt breeding, though one-year-old females of course can. Thus, there are now perhaps four times more breeding females available than males and the practice of polygyny has become not only advantageous but unavoidable.

Polygyny is only practicable where food is so plentiful that one adult alone can feed the young. It is interesting that two of the three great flightless birds of the grasslands—the rhea in South America and the ostrich in Africa—practice it normally, holding harems whose members lay in a communal nest, the eggs being incubated largely by the male who also cares for the chicks.

Emus in Australia are not normally polygynous (though occasionally more than one female will lay in the same nest) but the males are again responsible for incubation and care of the young.

These huge birds can avoid most predators once they reach maturity. Their running powers are prodigious—ostriches can exceed 40 miles per hour (64 kilometres per hour)—and they have height and keen vision to detect danger far off. Their eyes are set on the sides of their heads to give nearly all-round vision. This is, in fact, the case with most birds, whatever their size. Seedeaters can see through about 340° without moving their heads and they have binocular vision in a small arc of about 25° to the front. Thus they are able to examine their food and judge pecking distances as well as keeping a look out for danger at the same time. Birds of prey,

Left
Though a capable flyer and related to the birds of prey, the secretary bird is a terrestrial hunter. Standing over 3 feet (0·9 metre) high it strides across the grassland picking up insects and other small creatures. Larger prey is kicked to death and the species is skilful at killing snakes.

Above right
Bee-eaters often travel with, and ride on, large herbivorous mammals and even birds, swooping out to capture the insects that they flush from the grass. Here an African carmine bee-eater takes rapid flight from the back of a kori bustard.

Below right
A male pin-tailed whydah. Whydahs are brood parasites, laying their eggs in the nests of waxbills. Some always choose the same host species, mimicking both its eggs and young so their own chicks are raised together with its.

Left
In what is a remarkable example of convergent evolution, the American eastern meadowlark (*below left*) and the western meadowlark (*left*) closely resemble in habits and plumage the African yellow-throated long-claw (*below right*) which is not a lark but a pipit.

Facing page
The banded plover occurs in grassland and open habitats in central and southern Australia, favouring areas close to water. When the chicks are hatched, they leave the nest at once and the parents will carry out 'distraction displays' feigning injury in order to lure possible predators away.

on the other hand, have far fewer enemies but need good forward vision for their hunting so their eyes are set towards the front of their heads giving a total arc of 120° or less but binocular vision extending through 80°.

Raptors are generally common in the grasslands. On the ground, the long-legged secretary bird hunts insects, small mammals and reptiles, killing the larger prey by strong blows with its feet. In the air, falcons hunt other birds. Broad-winged birds of prey soar on the thermals that rise from the heated ground, seeking both live prey and carrion. The presence of great numbers of herbivores, such as zebra and wildebeest, means ample food for vultures. Soaring high, they scan a wide area and when one descends, others see and follow. In Africa, the first to arrive at a carcass will usually be Rüppell's griffon or white-backed vultures. They cannot break the hide of large animals but often this has already been done by lions or other animals. Their heads and necks are bare so that they can thrust them inside the carcass without fouling their plumage. Lappet-faced vultures have strong beaks and feathered necks; they can tear the skin and eat hard, outer muscle. On the fringe of the flock, the smaller Egyptian vultures wait to pick the scraps left when the other species have finished. Generally, vultures hunt by sight. Only American turkey vultures, which have exceptionally large nostrils and olfactory organs, are known to detect carcasses by scent.

The big herbivores of the grasslands are not only useful to birds when they die. Flycatchers, such as the South American cattle tyrant, will move with grazing animals, often riding on their backs, to catch the insects that associate with them or are disturbed from the grass by them. Cattle egrets feed on insects, lizards and frogs flushed by grazing animals and will also pick ticks, leeches and other parasites from the animals themselves so that the relationship is of mutual benefit. The spread of stock farming has provided them with a new opportunity and in the last forty years they have spread from their original range in Africa and Asia to North and South America, Indonesia, Australia and New Zealand. The African oxpeckers, distantly related to starlings which may also be seen to adopt the habit occasionally, feed exclusively on the external parasites of large mammals and are built rather like woodpeckers, with sharp claws to hang on and a stiffened tail to give support.

Birds may also associate with other creatures for safety. The black-headed weaver will sometimes build its nest in close proximity to the

Left
The red-billed firefinch is a widespread and common bird of wooded African grassland, feeding largely on the small seeds of grasses, whose leaves it uses to build an untidy woven nest, lined with feathers and usually located low in bushes.

Above
The division between grassland and desert is by no means clear cut in many areas, at least so far as birds are concerned. Related to other insectivorous grass warblers, the Damaraland rock-jumper is a little-known warbler adapted to life in the dry, rocky and shrub ground country of parts of Angola and south-west Africa.

dwelling of some dangerous beast such as a bird of prey, a man or wasps, and gain protection from other predators thereby.

Some African honeyguide species have developed a symbiotic relationship with the badger-like ratel. The bird follows bees back to their nest and then sings and flys excitedly until it attracts the attention of a ratel which, recognizing the meaning of this behaviour, follows the bird to the bees' nest and digs it out to feed on the grubs and honey. After it has finished, the honeyguide feasts on the remnants, even being able to digest the wax.

Honeyguides are brood parasites, laying their eggs in the nests of bee-eaters, barbets and other birds. When the young honeyguide hatches, it kills other chicks in the nest with sharp hooks which it bears temporarily on the tip of its beak for this purpose. Many other birds have adopted brood parasitism for the advantages it brings. A female can have more young in this way than she could cope with by her own efforts and, in a temperate area where the peak nesting season may be short, it is valuable for adults of a migratory species to be able to leave before there is any risk to themselves from food shortage.

Below
Found in temperate grassland in Eurasia, the little bustard is migratory at the eastern end of its range, moving west to avoid the cold winters of the steppes. The ploughing of the Russian grasslands has caused its decline. Here a female is incubating her eggs in the hot sunshine.

Left
Perched on a rhinoceros, two yellow-billed oxpeckers resemble woodpeckers on a tree as they hang on by their claws and lean back on their stiffened tails. The stout beak is used to extract external parasites such as ticks from the skin of the larger herbivores.

Below
Living in the east African acacia grassland, the superb starling breeds in two rainy seasons. Feeding particularly on termites and ants, they start foraging in the early morning before the heat of the sun drives their prey back into its inaccessible nests.

Not all the birds that are parasitized by one species or another are helpless to cope with it. They may react by deserting, or by building a new nest on top of the old one and immuring the offending egg. Because of this, many brood parasites specialize in a single host whose eggs and chicks are mimicked by their own. With most species of African whydah, each uses one particular waxbill species to rear its young. As the chicks of each waxbill species have a specific pattern of markings inside the back of their throats, and no waxbill will feed any chick that does not bear such markings, young whydahs mimic these exactly. Being able to compete on equal terms with the other chicks in the nest, they have no need to eject or kill them and all are raised together.

So far as the relationship between grassland birds and man is concerned, the picture is mixed. The clearance of forests for agriculture has extended a quasi-grassland habitat over much of eastern America and western Europe, so that some species have benefited. Elsewhere, much

natural grassland has lost its wide variety of wild herbivores and these have been replaced by domestic livestock. This process has usually been accompanied by extermination campaigns against predatory mammals and some birds, notably raptors, more often than not in determined ignorance of their real biological roles. The effects on birds have been greatest where grassland has been ploughed up for agriculture. Some species have lost the niches on which they depend. Others have been hit by pesticides and herbicides, more because of the reduction in natural plant and insect foods than through direct toxic effects, though these have affected species in some areas significantly. A few species have benefited greatly from man's crops and predator control, becoming major pests that defy all efforts to reduce them.

Deserts

Deserts cover one fifth of the world's land surface. They stretch across North Africa and the Middle East into central Asia, and through the south-west of North America into Mexico; they occur in south-west South America and south-west Africa and they occupy the centre of Australia. Rainfall is rare and often irregular – for example, it rains in the Sahara every year but not usually in the same place twice in succession. When it does come, the entire supply for several years may fall in a fierce storm of just a few hours duration. High winds, on the other hand, are frequent, often carrying sand or dust and contributing to the drying out of the land.

Depending on their geographical locations, different deserts have different temperature ranges but daytime temperatures are generally high and in some areas the surface of the ground may become as hot as 80°C. Equally significant is the great drop in temperature that occurs at night. In most areas the ground is shielded neither by vegetation nor moist air masses and it radiates heat rapidly, its temperature falling to near freezing even in subtropical regions.

Humidity is generally very low but the deserts of south-west Africa and south-west South America are influenced by the cold ocean currents which cause fogs, thus preventing sunlight from reaching the ground and reducing the temperature range.

Right
Zebra finches, unlike most related birds, but in common with other species that live in the dry centre of Australia, have developed the ability to suck up water without raising their heads and thus to minimize the time that they must spend at water holes where predators concentrate.

Below
The stone curlew nests in central and southern Europe, North Africa, south-west Asia and India, usually in open habitats varying from arable farmland to semi-desert. Feeding primarily on slugs, snails and worms, the birds are most active at night when their prey is itself on the move.

Above
A female pin-tailed
sandgrouse on her nest.
Breeding in Iberia, North
Africa and eastward to
Afghanistan, it is not
exclusively a desert species
but also occurs in other dry
areas. The belly feathers of
sandgrouse will hold a large
volume of water and thus it
can be transported to chicks
located far from a drinking
place.

Left
The cactus wren is a bird of
arid areas in southern North
America. Each individual
builds a number of spherical
nests, usually sited in the
most prickly part of a cactus
or thorn bush, and uses them
to roost at night and as cold
weather shelters.

By no means all deserts are sandy. There are vast areas of clay, stone or rock and in some areas the ground is impregnated with salt. The vegetation is usually impoverished or absent because of the lack of water and the violent changes in temperature. Trees are few, stunted and thorny. Bushes are largely concentrated in the marginally more favourable areas–for example, where ravines or dry water courses offer some shelter and water may lie closer to the surface of the ground. Most plants are annuals, their seeds lying dormant in the soil perhaps for years, until the rains come and they can spring to brief life–growing, flowering and producing their seeds before the waters dry up and the landscape becomes baked and cracked once more.

Thus, food for the living creatures of the desert is usually very scarce and it becomes abundant only locally and more or less irregularly. As elsewhere, birds have risen to the challenge and solved here the problems of temperature, aridity and food shortage.

Probably less than half the species found in deserts are confined to the habitat, the remainder also occurring in other more benign situations where their habits may be different as a response to easier living. The avifauna of the Old World

deserts is better adapted than that in the Americas, where deserts have not existed for so long.

Birds are helped considerably by their naturally high internal temperatures to cope with major external fluctuations. It appears to be the heat of day rather than the cold of night which has most influence on them. Desert-living individuals of species which also occur in other habitats are smaller than those living in cooler areas, thus increasing the capability to radiate heat. Their plumages tend to be 'tight fitting' as this reduces the insulating effects. When in danger of overheating they pant, so that the flow

of air causes evaporation and consequent cooling. Advantage is taken of natural shade under plants or rock, particularly for the siting of nests, whose contents must be protected from the extreme heat of the mid-day sun. Perhaps it is for this reason that the Australian dotterel covers its eggs with bits of dried soil when it leaves them, though this also serves to hide them from predators. Surprisingly few species nest in burrows or holes though the temperature is much more equitable, even a few inches below the ground surface.

Many desert birds are pale-coloured and this seems to have several advantages. Pale colours reflect heat; they require less of the bird's energy

Above left
In the intense heat, the feathers of this Indian courser are fluffed out and the bird opens its beak to pant so that internal moisture is evaporated and its temperature is lowered. Coursers will follow flocks of large domestic livestock in order to take the insects they disturb.

Above
The lanner is a large African falcon widely distributed in open country. Its range extends well into the Sahara and covers the Kalahari and Arabian deserts. Lanners feed on small birds caught in flight and will also take mammals and insects including locusts.

Right
When an Australian dotterel
leaves its eggs in order to
feed, the bird covers them
up with a layer of dried earth
or clay, perhaps as a
protection both against
predators and the direct heat
of the sun. This species
inhabits the dry interior and
often occurs far from water.

resources to create than do dark pigments; they provide camouflage and to employ it to best advantage many prey species with such plumages crouch immobile when a predator approaches so that neither movement nor a shadow reveals them. The cryptic value of plumage is well demonstrated by the desert lark. It has a wide range across North Africa, Arabia and into Asia, living in rocky areas which themselves differ considerably in colour from place to place. Local populations of this bird match the natural colour of the land they frequent.

However, not all desert birds are camouflaged and some have striking black and white patterns. This is particularly noticeable in wheatears, where only the male has this coloration. Experiments have shown that the members of some wheatear species taste unpleasant. A predator has only to kill one or two individuals for the warning coloration to be remembered in future, so the birds make no attempt to conceal themselves, instead behaving so as to advertise their distinctive patterns. The females tend to be much duller–frequently pale brown and white–which may be necessary because they might otherwise draw attention to vulnerable eggs and young. Both sexes and all species have

very similar, simple, black or brown and white patterns on the tail and rump which are particularly evident in flight and act as a wheatear 'trademark'. However, not all species are distasteful and that which breeds in Britain was regarded as a table delicacy up to the turn of the century.

The ability to tolerate considerable dehydration is particularly marked in birds of arid regions. This is important not only because water may be difficult to obtain but also because cooling, particularly by panting, uses up much of a bird's body fluids. Some species may be able to lose up to half of their weight through dehydration and still recover. Many or all can rehydrate themselves very rapidly. Mourning doves that have lost about a fifth of their weight can replace it in a single ten minute drinking bout.

Seedeaters need to visit water regularly because their food is so dry. As a result, they either have to stay close to supplies or to be strong fliers. Sandgrouse will make a daily round trip of up to 60 miles (96 kilometres) to drink. By bathing and saturating their belly feathers, which have a special structure that retains water, the males can carry adequate supplies back to

their young, which extract it with their beaks. Other seedeaters feed their young on insects and these contain adequate water. Pigeons supply their chicks with 'milk'—a secretion produced in the bird's crop and much like rabbit's milk in its protein and fat content.

Because there are few sources of water and birds concentrate at them in large numbers, birds of prey also frequent these sites so drinking is risky. It is necessary to approach cautiously and pigeons, amongst others, suck up the water rapidly without raising their heads, so that they need not delay. Most birds in other habitats lift their heads between each beakful and allow the water to flow down the throat.

Species that feed on green vegetable stuffs, on insects or on other animals find water in their food, and some may not need to drink at all.

It was in a species of desert-living nightjar, the American poor-will, that true hibernation in birds was first confirmed by ornithology, though it had probably long been known to the native Hopi Indians. Research has shown that the poor-will's normal temperature fluctuates

Above
A mourning dove nesting in a prickly pear. This species is widely distributed from southern Canada to Central America, occurring in most habitats including arid zones. The chicks are fed on 'pigeon's milk', a protein-rich secretion from the adults' crop, and can thus survive in conditions where water or food with a high liquid content is not available.

Right
The tawny pipit breeds in most open habitats over much of Europe, central Asia and North Africa, migrating in winter. Its long legs and slender build are suited to pursuing insect food on the ground and are reminiscent of the related wagtails, though the latter are associated with wetland habitats.

considerably, unlike most birds, and this increases its tolerance of extreme heat. When resting normally its metabolic rate drops to half or less than that of other birds of similar size, so that it needs correspondingly less food. When winter temperatures are low and the bird's insect prey is not on the wing, its body temperature may be greatly lowered even to just above freezing point, its heart beat is slowed and it becomes fully torpid, using its stored energy

reserves at only a fraction of the usual rate. When ambient temperatures rise, and insects are again active, the poor-will quickly reverts to its normal conditions. So far, no other bird species has been shown to have such a well-developed ability though several, including nightjars, humming-birds and swifts, can become torpid for shorter periods.

Birds are generally very thinly distributed in deserts. With few or no trees, most species spend much of their time on the ground and have developed good running ability. The roadrunner, a North American cuckoo that preys on lizards, other reptiles and invertebrates, is a bantam-sized bird with a 10 inch (25 centimetre) tail that can run astonishingly quickly and has been timed at 18 miles per hour (29 kilometres per hour). Its long tail acts as a rudder to give it considerable manoeuvrability so that it can evade most ground predators without deigning to fly. Nonetheless, for most desert birds, good flying abilities are essential in order to reach the thinly scattered resources of food and water.

Breeding is related to rainfall rather than season. Particularly in Australia, where a number of wildfowl species breed in the central zone, many species are nomadic and will move from area to area in pursuit of good conditions. Where rains are more reliable birds may migrate into an area to breed but move out again as soon as the brief effect of the rains has passed. In very dry years, many birds may not breed at all and so populations of desert species fluctuate considerably.

When rain does at last come, no time can be lost. It is thought that most nomadic species mate for life so that as soon as the rains begin, breeding can commence without delay. The urgency is so great that courtship may start within minutes of the first raindrops falling and in a few hours birds will be building their nests. Thus the plant growth triggered by the rain will reach its peak production of food at the same time that the young are hatched.

The special conditions of the desert affect the breeding birds in other ways. Australian zebra finches outside the arid zone will hold nesting territories one to a tree but in areas where trees are scarce, several may share. Indeed it may be unprofitable for the male to spend time defending

territory when he can help the female with nest construction thus reducing by several days the time before eggs can be laid.

A few rather unexpected birds use the desert habitats. The grey gull of Ecuador, Peru and Chile feeds along the Pacific coast, particularly taking small crustaceans from the sandy beaches, but it breeds inland on the arid deserts that lie between the coast ranges and the Andean chain. The nest is a scrape in the ground and at night, when temperatures fall to near freezing point, the gulls incubate the eggs or brood their young in a normal manner. In the early morning a wind springs up, at first blowing off the cool land mass to the warmer sea and later, as temperatures rise,

blowing on to the land from the sea. It prevents air and ground temperatures from rising too high. Nonetheless the heat is considerable so each gull stands over its nest facing into the wind, allowing the moving air to help cool the contents and also shading them from the direct sun. For a seabird species to undertake breeding in such an unlikely environment there must be some special advantage gained and it seems possible that the comparative scarcity of predators in the arid desert compensates for its difficulties.

Vast numbers of other birds of many species must venture into the deserts twice yearly. In Africa and Asia, the great band of desert forms a major hurdle which migrants have to cross as they travel between northern breeding grounds and their southern wintering areas. Vast numbers perish and as the Sahara continues to grow in size the journey must become that little bit harder every year.

In most habitats, man's activities have destroyed vast areas and continue to change them. Paradoxically, many desert species may have benefited from the land management, or mismanagement, practices that created deserts in the Middle East in prehistoric times, in the USA much more recently and are still developing them south of the Sahara today.

Top
A greater sand plover settling on its nest. The species breeds in the deserts and steppes of Asia often far from water, its food varying with the habitat but basically a wide range of invertebrates. Outside the nesting season the birds frequent coastal habitats stretching from South Africa to Australia.

Above
The pyrrhuloxia occurs in dry mesquite scrubland on the borders of Mexico and the USA. Related to the cardinals and, more distantly, to the buntings of Europe, it feeds mainly on seeds and berries but also takes insects. The nest is built fairly high in a thorn bush.

Mountains

For every 500 feet (152 metres) that the ground rises, the temperature drops 1°C, so that increasing altitude creates progressively greater cold. At the same time the wind becomes stronger and blows far more frequently. The soil gets shallower and poorer, the humus that is produced by the plants being washed or blown to lower levels. The character of tree growth changes with increasing altitude. In the tropics, rainforest is replaced by deciduous woodland and then by conifers until the tree line is reached; above it only shrubby plants can survive and they in turn are succeeded by alpine grasslands. At the highest altitudes, the cold and lack of soil defeat even grasses, and only mosses and lichens survive, forming a tundra-like vegetation up to the snowline where polar conditions prevail, but without the rich feeding that polar seas provide.

This vertical layering of habitats by altitude parallels their latitudinal distribution between equator and poles. What is most striking is that they are condensed into such a short distance. At the equator, a 20 000 foot (6 000 metre) mountain contains a range of habitats that you would have to travel about 6 000 miles (9 000 kilometres) to find if you remained at sea level. At latitudes further away from the equator the tree line and the snow line occur at increasingly

Left
Verreaux's eagle breeds in Ethiopia and southern Africa, frequenting broken country and mountains up to high altitudes, and feeding primarily on hyraxes which live in colonies on the rocky ground. The nest is almost invariably located on a cliff ledge and the sites are traditional.

lower altitudes. In much of Britain, trees naturally grow up to about the 2 000 foot (600 metre) contour, though in many places they have long since been felled and sheep farming has prevented their return. Snow may lie throughout the year in shaded corries at about 4 000 feet (1 200 metres).

As a result of the presence of these distinct habitats in close proximity, mountains often contain different bird communities close to each other. At the higher levels, they often show well the ways in which species evolve to occupy vacant niches. For instance, mountains in tropical areas may carry temperate and subarctic habitat types but because they are isolated from the main geographical zones in which these habitats exist, the bird species that would normally be present there do not occur. Instead, tropical species from the surrounding lowlands,

taking advantage of the lack of competition, have evolved forms able to tolerate the non-tropical conditions. Ultimately these forms may become separate species in their own right and, because their ancestry and interrelationships are usually clearer than those of birds in less isolated habitats, they are of particular interest to students of evolution.

However, it is by no means always the case that birds on mountains are related to the lowland species. At least in temperate areas some–such as ptarmigan–may be relict populations of species which today have a more northern distribution. Failing to follow when the ice sheets last retreated, these outliers instead ascended into the mountains and were marooned there by the rising temperature in the lowlands.

Though the habitats generally resemble those found elsewhere, mountain life does impose some

special demands on birds. At higher altitudes winds are generally strong, so that most birds are either large species that can fly powerfully—particularly birds of prey such as vultures and eagles—or small ones that keep close to the ground where they can find shelter.

Above about 16 000 feet (4 800 metres) the thinness of the air becomes a problem. Flight presumably needs more effort, and the effort needs more oxygen, which is itself less easy to obtain. Studies in North America of species which occur in both highland and lowland zones have shown that individuals from high altitudes have hearts at least ten percent heavier than those from areas where air density is normal. This is assumed to aid blood circulation and thus the transport of oxygen to the tissues. Similarly, montane birds are bigger than their lowland counterparts to reduce heat loss.

Above
Griffon vultures are the commonest large European vultures and are found in highland areas (and hunting over the surrounding land) around the Mediterranean and east to India. When the first birds descend to a carcass, others are quickly attracted from further afield and many will soon congregate, feeding to repletion if undisturbed.

Right
The chukar partridge and the similar rock partridge both occur in barren upland areas, the former from Turkey to central Asia and the rock partridge in south-east Europe. Both species live on the ground, feeding on seeds and other plant material as well as insects and other invertebrates.

Below
A juvenile broad-tailed hummingbird perches on a pine twig in the Rockies. Like other upland hummers, they take insects as well as feeding on nectar. When temperatures fall at night, they become torpid, reducing the metabolic rate to conserve energy. This species winters in Mexico.

Above
The dotterel breeds close to the snow line in high tundra and mountain ranges in Europe and Asia, moving south to the Mediterranean and Middle East in winter. There is also a small atypical population recently established on the Dutch polders, below sea level. The eggs are incubated by the male who also looks after the chicks.

Several hummingbird species live in montane habitats in North or South America, where they tend to eat insects at least as much as nectar. One of these, the giant hummer which occurs in the Andes, is the largest species of all, being about the size of a starling. Most, however, are much smaller and face a serious heat loss problem at night when the temperatures are lowest. To overcome this the birds become torpid when they roost; their body temperature drops and metabolism slows down. This ability is very uncommon in birds. Most species have sufficient size and insulation to make it unnecessary or have found other ways of solving the problem, such as migrating away from unfavourable

The rock wren is confined
to the alpine zone of South
Island, New Zealand. Very
poor flyers, the birds live in
areas of tumbled rock,
spending much of their time
out of sight, where they are
protected from snowfall and
extremes of temperature and
able to seek their insect food
at all seasons.

Above
The Alpine swift nests
colonially in cracks in cliff
faces or artificial sites such
as tall buildings. Swifts feed
entirely on insects caught on
the wing and may fly a total
of over 600 miles (960
kilometres) a day when
feeding young. This is one
of the largest species, with
a wingspan of nearly 2 feet
(0·6 metre) and a top flying
speed in excess of 100 miles
per hour (160 kilometres per
hour).

Right
A male rock thrush foraging
for insects and berries or
small reptiles. Breeding
across southern Europe and
central Asia, the species
winters in Africa. The male
holds two territories—a rocky
one for nesting and a more
grassy area for feeding.

conditions. Torpidity does not occur in the incubating females, which must maintain the temperatures of the eggs or young. They are themselves to some extent insulated by the nest which is deeper and thicker than those of tropical hummers.

Montane birds site their nests with particular care because of the extremes of temperature. Some species that build exposed nests choose sites that are warmed by the early morning and evening sun but shaded from it during the middle of the day when, striking down through a clear sky and thin atmosphere, its heat may be excessive. Shelter from the wind is also important.

Many birds, including raptors, ducks, woodpeckers, pipits and flycatchers, nest in natural holes in scree, under boulders or in cliffs. Others excavate their own nest tunnels with beak and feet or share burrows excavated by rodents—snow finches living with the rabbit-like pikas in the Himalayas and little owls with marmots in the Alps. Thus they find shelter from the wind and fairly constant temperatures because the rock or soil mass around them loses its warmth only slowly at night. Often in the montane situation the nest site is also used as a nightly roosting place throughout the rest of the year. However, some species, notably finches,

roost communally; over 200 birds have been recorded in a single rock cleft, huddled together to diminish heat loss. Other species avoid the cold altogether by flying down to the warmer valleys and many mountain birds descend to lower altitudes in the winter–wall creepers in the Alps, river chats in the Himalayas and juncos in the Rockies need to migrate only a few miles to enter milder habitats.

Mountain ranges form barriers. Like oceans they tend to separate the terrestrial birds that live on either side of them and particularly to split migration routes for smaller species that find it easier to travel around than over them.

Presenting problems for human occupation as much as they do for birds, at least the higher reaches of mountain systems remain relatively unspoilt and unthreatened except in a very few areas where tourist pressures are excessive.

Left
The snow finch is found in mountain ranges from Spain to the Himalayas, remaining above the tree line and up to the edge of the snows for as long as weather permits. The nest is placed underground in a crevice or an animal burrow.

Above
Townsend's solitaire is a thrush that breeds at high altitudes in and above the scrub woodlands of the Rockies from Alaska to California. Birds from the northern half of the range move south in winter. Like many other species, insects are taken extensively in summer but berries are important at other seasons.

Wetlands

There are many types of wetland. Inland, there are various freshwater habitats – upland streams and great rivers, tundra lakes and tropical swamps, mountain bogs and lowland reedbeds. Some of these are only wetlands periodically – for example, the tundra freezes in winter and the wetlands of the Australian interior may be dry for years on end until the rains come. There are the brackish areas – both inland lakes in hot regions where a high evaporation rate makes the water exceptionally salty, and estuaries where fresh and salt water meet and mix. On estuaries and elsewhere along the coast, salt marshes may

form, their vegetation and invertebrate life influenced by the sea on one side and by fresh water on the other. Thus there is no clear cut division between the freshwater and the marine environment. The International Waterfowl Research Bureau has drawn an arbitrary boundary defining the 'lower limit' of wetlands as 6 metres (20 feet) below the tideline but many wetland birds move beyond this into deeper seas.

Because of their diversity, wetlands hold a large variety of birds. In uplands, watercourses are usually fast flowing and rather poor in nutrients. Little vegetation can grow and insects

Left
Unlike most fishing species, the anhinga or snake bird does not grasp prey in its beak but spears it. With a very heavy bone structure and poorly waterproofed plumage, it swims low in the water and can dive easily but has to come to land to dry out its feathers.

Right
The common sandpiper is widely distributed in summer along rivers and around lake shores throughout boreal and temperate Eurasia, finding invertebrate food in the waterside vegetation and in the shallows. In winter it migrates to Africa, south-east Asia and Australia. A closely related species, the spotted sandpiper, breeds in North America and also migrates south in winter.

and invertebrates seek anchorage and protection
beneath the stones. Only a few species, such as
torrent duck and dipper, can flourish there. As
the land levels out, rivers flow more slowly and
gather more nutrients so that more plants grow,
there is a greater variety of animal food and of
birds—for instance kingfishers and dabbling
ducks such as teal may live here. In their lower
reaches, rivers may meander over wide areas of
marsh, which are ideal for members of the heron
family.

When it reaches the sea, the river estuary may
be a zone of exceptional fertility, supporting large
numbers of breeding waders, gulls and terns. The

Above
The long spatulate beak of
the spoonbill is moved to
and fro beneath the surface
of shallow water, sifting out
small swimming creatures.
The different feeding
adaptations of different
species of large wading birds
prevent competition even
when they are feeding on
similar foods.

Left
The little gull winters in
coastal waters but moves
inland to breed on grassy
marshes and flooded
grassland where there is
ample emergent vegetation.
It feeds by picking fish and
insect food from the water
surface in flight and by
catching insects on the
wing.

Left
Since it is quite conspicuous, the oystercatcher discreetly leaves its nest when danger appears. The eggs or chicks, being camouflaged, are hard to find. In some areas it is a coastal breeder feeding on shellfish, but elsewhere it breeds inland on damp moorland or grassland and feeds on earthworms and other land invertebrates.

Below
Because they have evolved so slowly, feather lice may give clues as to the relationships of birds. It appears from them that flamingos are more closely akin to ducks than to such birds as storks and herons. The greater flamingo's beak is designed to filter small aquatic animals and algae from the saline lakes in which it feeds.

whole seashore is potentially a nesting place for other waders, ducks and seabirds. In winter, an even greater population of species that breed in the tundra and moorlands of the north migrate to the milder estuaries and inshore waters.

Tundra and upland bogs constitute a vast wetland zone, particularly in the northern hemisphere where wildfowl and waders breed in great numbers, though often at low densities. Upland lakes are usually deep, cold and nutrient-poor, supporting a few fish-eating birds such as goosander. However, lowland lakes are warmer and richer. Diving ducks feed in the deep water, surface feeders in the shallows, herons on the margins. Reed warblers are characteristic of the interface between reedbeds and willow, nesting in the former and finding their insect food in the latter.

The variety of wetland habitats and the large number of niches available has allowed many different species to evolve. However, the basic choice is between being a swimming bird or a wading bird and so most species, whatever their evolutionary origins, conform to one of two shapes–'duck' or 'wader'.

The 'duck' shape of divers, grebes, wildfowl and others is suited to swimming on or beneath the water. The body is streamlined and clothed with waterproof plumage, the structure of the feathers allowing them to interlock and repel moisture. The legs are generally placed well to the rear of the bird for efficient propulsion and this tends to make it more or less ungainly on land. The feet of many species are webbed. When the bird

Above right
The grey wagtail is found mainly by fast-flowing fresh water, usually in upland areas. It hunts for insects both by running on the ground and by flycatching in the air. The nest is usually built in a cavity and the chicks are fed there until they fledge.

Below right
Reed warblers suspend their neatly woven nest from the stems of reeds, usually over water, but feed on insects collected outside the nesting habitat in adjoining willow trees. Breeding in Europe and western Asia, they winter in the reedbeds, grassland and savanna of Africa.

Right
Seen in flight, a dipper shows its rather short wings, which it uses to swim underwater in fast flowing streams, where it searches on the bottom for aquatic insects. The dipper is the only passerine species to have adapted to feeding underwater.

brings its foot forward through the water, the toes come together so that the webs fold and the foot turns back to be pulled through the water with minimal resistance: on the back stroke the whole foot unfolds and becomes a rigid paddle pushing against the water. The toes of coot, which belong to the rail family, are not joined by webs but have lobed sides. These are less efficient for swimming but are better on land where coots spend some time foraging for food.

Competition between species for food is prevented by minor but significant behavioural and physical adaptations. Ducks demonstrate this adaptive radiation very well. They can be divided first into two general groups—surface feeders and diving ducks. The first group find their food, which consists of plant matter, seeds and small creatures, on the water or down as far as they can reach by up-ending. Mallard up-end frequently and so do pintail which, having longer necks, can feed in deeper water. Teal, being smaller birds with shorter necks, feed in the shallowest zones, mainly on small seeds in winter. Shoveler have spatulate beaks which are modified to act as strainers so that by sucking in water at the front and expelling it at the sides of the bill they can extract any food material. Wigeon largely feed on the land near water, grazing

Above
The bearded tit breeds in large reedbeds. The nest is suspended amongst the reeds, often close to the ground or water level, and the young are fed on insects. Outside the breeding season the main food is small seeds or marsh plants. Vulnerable to hard weather, the European population may be considerably reduced by a cold winter.

Right
The yellow-headed blackbird is a member of the large American family of icterids and nests exclusively in marshlands in western North America, especially on the prairies. The nest is suspended over the water and the young are reared on large insects. Breeding colonies may hold as many as 200 000 nests.

grasses, as geese do. Because they do not dive, surface feeders are found mainly in shallow water areas characteristic of marshland.

The diving ducks seek their food beneath the surface, using their feet for swimming. Some, such as pochard, take aquatic vegetation; others feed on molluscs and others catch fish. The mergansers are particularly well equipped for grasping slippery prey. Their long bills have hooked tips and rows of small knobs along the edges, rather like teeth, and there are horny bristles on their tongues.

The 'wading' birds such as herons, storks and waders are species which frequent water margins. They generally have long legs to keep their body

plumage dry and long necks and beaks to give
them extensive reach under water or into mud.
They usually have long toes which combine good
mobility on land with support on soft surfaces
such as floating vegetation. Phalaropes, which
habitually swim, have partially webbed feet.

Wading birds feed characteristically on animal
prey. Herons will take any live prey that they
can manage to capture; the grey heron feeds not
only on fish and frogs but also takes water voles
and birds up to the size of a dabchick. The typical
hunting mode is to stand motionless or move
stealthily until prey comes within striking
distance, when the long neck shoots out and the
beak grasps it firmly. Herons vary considerably

in size, the goliath heron of tropical Africa
standing five feet tall and the little bittern hardly
more than one foot; its much smaller size allows
it to hunt small prey in dense cover and it can
clamber about skilfully in reeds and bushes.

Waders proper feed almost entirely on
invertebrates, such as insects and their larvae,
worms, small crustacea and shellfish. Some
species have short bills and pick their food off
the ground or, like dunlin, feed on small snails
that live just beneath the surface of the mud.
Those with longer beaks, like redshank, take
somewhat larger prey that burrows deeper, while
those with largest bills can probe down in pursuit
of such large and deep burrowing animals as

ragworms. Hunting techniques differ too. Dunlin probe rapidly at random, relying on the speed and thoroughness of their search to produce results. Grey plover stand motionless to watch for the telltale movement of sand that betrays the food beneath and then run forward to grasp it. Species such as oystercatcher take large shellfish. Opening the various types of shell demands different but equally skilful techniques. Different individuals usually specialize in feeding on either mussels or cockles.

Many waterbirds draw little or no distinction between fresh, brackish or salt-water habitats for their feeding. Waders, such as ringed plover, breed mostly on sandy or shingle beaches but some have recently begun to colonize inland habitats, notably gravel pits. Dunlin breed on wet moors, marshes and tundra, wintering on brackish estuaries and salt-water coasts.

The muddy expanses of estuaries hold great densities of invertebrates on which waders feed, often in huge numbers. Their activities are governed by the movement of the tides. When it is high water, none can feed and they form dense roosting flocks at some location above the tide limit on salt marsh or open ground where they can see danger approaching. As the tide begins to fall and wet mud is exposed, the first birds leave the roost. These are the smaller species that run behind the retreating wavelets picking the small food items near the surface of the mud. When the water has retreated further and exposed the habitat of larger invertebrates that live further down the shore, other wader species move out in turn. After all the other birds have left the roost, some oystercatchers may still

remain, waiting until almost low water when the mussel beds on which they feed will be exposed. It is very apparent that waders have a time sense. Even when roosting out of sight of the sea, perhaps behind a sea wall, they know when to move out to feed.

In good weather waders may only need to seek their food on daytime tides but they may have to feed both by day and night when conditions are cold. Then their prey burrows deeper and is less active so that it becomes harder to find: at the same time they need to consume more to compensate for extra heat loss. In very cold

Above
Needing plentiful food to support their large size and large colonies, and sensitive to disturbance, the white pelican now has a scattered distribution in south-east Europe, southern Asia and Africa. Usually two eggs are laid and the chicks are fed initially on partly digested matter that the parents regurgitate, later on fish carried back in the pouch.

Right
Many large herons are colonial but the bittern tends to be solitary, its cryptic plumage concealing it in the reedbeds it inhabits. When alarmed, it stretches itself upwards, to look as much like part of the background as possible. The eyes are set low in the head so that it can watch what is going on at the same time.

weather, vast numbers of birds may be forced to make emergency movements over long distances in search of better conditions.

Wildfowl and some other water birds, such as grebes and divers, also use estuarine and coastal waters. Some visit them only briefly or when conditions inland are bad. The sea retains its warmth in winter longer than the land so coastal weather is often milder than inland and salt water freezes at lower temperatures than does fresh. However, ingested salt causes birds to dehydrate and some species cannot live in a saline habitat for long. Some wildfowl, like seabirds, have special glands which will extract the salt from their body fluids. Mallard are freshwater dabbling ducks and though they may roost on the sea do not normally feed there. Those that breed on the tundra usually migrate south in winter but the

Greenland subspecies instead takes to the sea, where it feeds on seaweed. To do this it has both reverted to diving and developed a salt gland like that of true seaducks.

All seaducks dive and most feed on molluscs or fish. The long-tailed duck can dive to depths of over 60 feet (18 metres) or more and stay submerged for several minutes. Like waders, seaduck often occur in dense flocks outside the breeding season, congregating where food is prolific. They can usually feed at all states of the tide and they roost at night on the water, often moving away from the shore, probably to areas where surface currents are less strong and a risk of 'wreck' in the event of storms is reduced.

Many water birds nest on or near the ground because they live in habitats where there is no alternative. Waders in particular often nest in

very exposed situations. Many plovers make only a simple scrape in the open ground and lay camouflaged eggs, usually having dark brown markings on buff, which are extremely hard to see. As there is no nest structure to draw attention, only the presence of the incubating bird is likely to lead a predator to them, so as soon as a predator is seen approaching, the parent runs off to some distance from the nest. In open ground with few landmarks it is most unlikely that a predator will be able to mark the spot from which the bird came even if it was spotted as soon as it started to move. While the predator is in the area, and particularly if it approaches too close to the nest, one or both parents may harass it by calling and flying close, or carry out a distraction display, feigning injury and flopping over the ground to lure the predator away. This type of reaction and display are more

frequent and intense when the young are hatched.

Wildfowl, on the other hand, generally site their nests in some cover, perhaps beneath dense overhanging vegetation, and construct a deep bowl, usually lined with down plucked from their own bodies. This warm structure may be necessary because ducks lay more eggs than waders and a large clutch is harder to keep evenly warm during incubation. The eggs are light coloured – usually whitish blue or green. When a predator approaches, the cryptically plumaged female sits tight and only when her own life is in imminent danger will she flush from the nest.

In many wader species both parents take turns to incubate the eggs but in wildfowl the responsibility is often solely the female's as the brightly feathered male would attract attention. Indeed, in many species the male never sees nest, eggs or young, though he may continue to hold

Above left
The drake smew, like other wildfowl, comes into breeding plumage in winter. It starts courtship before the migration from the mild inland and coastal waters to the breeding grounds, which stretch across the Eurasian boreal forest belt, where smew nest in holes in trees.

Top
The dark-bellied brent goose breeds in Arctic Siberia and when the weather is poor, as may happen quite frequently, no young at all are reared but in good seasons the population may increase by as much as forty percent, thus balancing out the bad ones. Grey plovers, which also breed at very high latitudes, may also fluctuate in numbers.

Above
A small group of cattle egrets ride on the back of African buffalos. They feed on insects and small reptiles disturbed by the large animals as well as taking ticks and lice from their skins. With the spread of cattle breeding, since the 1930s cattle egrets have begun to colonize South and North America, Australia and New Zealand.

a waterside territory to which the female will come for feeding breaks while still incubating.

Eggs are usually laid at daily intervals but incubation does not start until the clutch is complete. Thus all chicks hatch within a few hours of one another. In both wildfowl and waders the young emerge already clothed in warm camouflaged down with their eyes open and their legs and feet well developed. As soon as they have dried, the whole brood can be rapidly led away from the nest. Most ground-nesting species have nidifugous young because the risk of predation is so high if chicks are fed for long at the nest.

Wildfowl lead their young to the water. In some species, notably shelduck, one or more adults may look after a creche containing the young of several broods. Wader chicks remain on land or at the water's edge with their parents. The young can feed themselves from the outset and the parents' duty is to keep them in suitable areas and protect them from danger. If an adult utters an alarm call the chicks may 'freeze' or huddle

Above
Feeding largely on aquatic plants obtained from the surface or by upending but not by diving, the gadwall primarily favours lakes in open steppe or prairie country in Eurasia and North America, migrating south in winter from the colder areas.

Right
Widely distributed across the temperate and boreal region of North America, the bald eagle feeds mainly on dead fish, supplemented by live prey and food stolen from ospreys. Formerly much persecuted, it is now protected by law but declining due to loss of habitat and former pesticide effects.

beneath the parent until all is safe again. The chicks of waterfowl may dive.

Mortality is high both during incubation and later. Gulls and skuas, mink and rats, pike and parasites all take their toll, along with exhaustion, exposure and starvation. If they survive, feathers rapidly replace down and within a few weeks the young are independent.

Some wildfowl and waders do not nest on the ground. Wood sandpipers will sometimes use the old nests of other birds in trees, and several ducks, including goosander, nest in tree holes, while shelduck prefer burrows. Nonetheless all have nidifugous young.

Birds which nest in relatively secure sites normally follow a different pattern. Herons in trees, storks on buildings and bitterns concealed in reedbeds feed their nidicolous young in the nest until they are well grown, feathered and able to fly. Bitterns find the security necessary for this because they live in reedbeds which give cover and are not easy for predators to penetrate. Other birds such as herons, egrets and ibises usually nest colonially in trees. In many instances, several species will occur together and such colonial nesting may be partly necessary because of the shortage of suitable sites in some wetland areas. Undoubtedly, large nests in trees are rather easy for predators to find and it must be an advantage if the forces of several pairs of birds can be combined in defending the breeding areas. Though this system of rearing the young may mean that fewer die, it also entirely excludes these birds from colonizing the tundra habitat which provides such a rich breeding environment for waders and wildfowl.

Not all ground-nesting birds have nidifugous young. Waterside species such as wagtails, which feed on insects caught by rapid running and in flight just above the ground, build carefully concealed nests on or near the ground. Being small, they can come and go with less likelihood of detection than larger birds could, and the young remain in the nest until they are fledged.

Wagtails are passerines, a relatively recent order in evolutionary terms, containing many

Left
A great white heron stalks through the water. Great numbers of these birds were killed at the turn of the century to provide the 'aigrette' plumes or 'ospreys' used in millinery. Feeding both inland and at the coast the species has an almost worldwide distribution in warm temperate and tropical regions.

Below
A secretive species, the water rail haunts the deep cover of reedbeds, swamps, overgrown ditches and other similar sites. With its very slender body it can slip easily through dense vegetation, its long toes supporting it on soft ground. It seeks small live prey, including birds, and some plant matter.

members and highly successful in most terrestrial habitats. They have, however, not been able to colonize aquatic situations which are almost entirely occupied by waterbird orders that have been evolving their special adaptations for much longer. The dippers are the one passerine family that has learnt to find its food under water. There are several species; all are dumpy bodied birds, wren-shaped but about twice the size, with short wings and rather long legs. They feed on small

Above left
The ringed plover's disruptive patterning makes it hard to see amongst stones and shingle on the seashore, riverbanks or the tundra, where it nests. Unlike many other small waders, the ringed plover does not probe at random for food but looks carefully for it on the surface of the ground.

Left
With a very scattered breeding distribution in Central America, Europe, Asia and Australia, it is likely that the gull-billed tern was once much commoner but that land drainage has destroyed many of the shallow water sites it favoured. Unlike other terns it feeds mostly on land.

Above
Standing 4 feet (1·2 metres) tall, the sandhill crane nests and roosts in wetland areas but moves out to feed also on adjoining tundra, prairie or farmland where it largely takes vegetable matter such as roots, berries and seeds, cereals and grasses plus small animals and insects.

crustaceans, insects and the eggs of fish gathered amongst the stones at the bottom of fast-flowing streams. Thus they are not found in lowland areas where the waters run slowly. Their wings are used for propulsion and direction while they are under water and the feet, which are not webbed, help by grasping and turning stones. Because they are small birds they are liable to lose body heat more rapidly than large ones and so they have particularly dense plumage for insulation against the cold water. It is probably because

of the heat loss problem that most water birds are medium-sized or large. Though late-comers and relatively little adapted to this situation, dippers have been able to exploit it because no other waterbirds have evolved to fill the niche within their range.

Members of a number of other orders have also learned that the aquatic environment provides a rich food resource. A number of birds of prey, such as bald eagles in America, Pel's fishing owl in Africa and ospreys worldwide, have learnt to catch fish by swooping down and grasping them in their talons when they come close to the surface. The Everglades kite is a medium-sized, large-winged tropical American raptor that feeds on big, strong-shelled water snails. Catching the snail when it crawls out on to emergent

Above
With its exceptionally long toes, the pheasant-tailed jacana of tropical Asia can walk easily over waterlily leaves or other floating vegetation. The female has several mates, laying a clutch of eggs for each of seven to ten males who then incubate and rear the young.

Right
Hunting its food in a manner reminiscent of a heron, the shoebill stork suddenly dips its enormous beak under the water to grab a large fish, frog, young crocodile or indeed any live prey small enough to subdue and swallow. The large beak is probably an adaptation to aid grasping prey in muddy African marshes.

vegetation, the kite simply carries it to a perch, waits until the snail emerges from its shell, grasps it with a deeply curved beak and then twists it out.

These snails also provide food for the limpkin. This is a long-legged and long-beaked bird obviously well adapted to its life in swampy areas. When it finds a snail it wedges it into a suitable crevice and uses its beak to force off the operculum which acts as a door, shutting the entrance to the shell when the snail withdraws in to it.

Both the Everglades kite and the limpkin live in American wetlands from Florida southward. Though their South American habitat is still relatively undisturbed, much of Florida's rich and strange swamp lands have been drained in recent years. This is the common fate of most wetlands in developed or developing countries. There are many reasons for this, some much less valid than others. Flood prevention, agricultural improve-

ment and the eradication of malarial mosquitoes are all desirable objectives but in most parts of the world the people who carry them out have no knowledge of their biological implications and many schemes go far beyond what is strictly necessary or ecologically acceptable.

Perhaps the worst problem that wetlands face is their public image as wet, mysterious and unpleasant wastelands suitable only for the location of airports or polluting industry, which nobody wants in his own locality. Estuaries have suffered particularly. Obviously ports and some industries can only be located close to water and some future loss must be accepted. However, estuaries are, in their natural state, the most fertile and productive habitats on earth. Perhaps the fact that they support birds which we find beautiful is less important than their role as breeding grounds for many of the fish we eat. Their continued destruction might be more serious than we yet realize.

Above
Avocets feed on small crustaceans in shallow brackish waters. The upturned end of the beak is swept to and fro beneath the surface and is held partly open so that as soon as prey is contacted it can be grasped.

Marine Environment

Two-thirds of the surface of the earth is covered by seas and oceans which vary much in character from area to area, as the land surface does. Even to birds which only skim the surface or penetrate a small way beneath it, the differences are marked and influence distribution and feeding habits.

The most productive oceanic areas are around the polar regions, where winds and currents combine to cause water to well up from the ocean bottom, carrying nutrients with them. This supports an abundance of plankton which in turn is preyed upon by a host of larger creatures—fish of all sizes, squid, seals, porpoises and larger whales, the commercial fishing fleets of many nations and great numbers of birds. Cold currents also well up off the west coasts of South America and South Africa, both of which also support great bird concentrations.

By contrast, the calm, clear, blue waters of tropical oceans hold little life. Here there are no forces to bring up nutrients to the sunlit surface where plant life, in this instance phytoplankton, can develop and support animals. Birds are very sparse except around islands where these ingredients can interact on a modest scale.

Within the marine environment there are fewer niches for birds to exploit than in terrestrial situations. As a result, there are less than 300

Left
The storm petrel also feeds on plankton, fluttering and pattering across the surface of the sea to pick up small crustaceans. This species is confined to the north-east Atlantic and Mediterranean and will dig its own burrow if no suitable crevice can be found.

seabird species in the world, compared with over 8300 land birds. With the exception of such notable long distance migrants as arctic tern and sooty shearwater there is relatively little movement of northern hemisphere birds to the south and vice versa, mainly because the intervening seas are so poor. As a result, each hemisphere has its own separate seabird community, developed partly from different ancestral stock. Comparison between the two is interesting because they show examples of convergent evolution. Though few, the different niches are extensive and several seabird species are far more numerous than any land bird.

Most seabirds spend a substantial amount of time on or under the water and need to be well insulated with fat and waterproof plumage. Flight necessitates a light body structure and this also gives buoyancy, which is a disadvantage to diving birds. They tend to have heavier bones and reduction of the air-sacs which are part of bird's respiratory system. The Galapagos cormorant, living where there were no predators, has abondoned the power of flight to become so lacking in buoyancy that it swims with its back awash. Other cormorants deliberately wet their

Left
The eggs of guillemots are strongly tapered to reduce the risk of their rolling off the narrow ledges. The Brunnich's guillemot (front bird) shows the thicker beak and slightly more robust build than the common guillemot (with 'bridle' around eye). Brunnich's guillemot has a slightly more northerly circumpolar distribution than the common guillemot.

Above
Immature herring gulls and a few adults congregate at food. Even more than most gulls, the species will eat a wide range of live and dead matter and occurs far inland, especially in North America, exploiting rubbish dumps and agricultural areas as well as sewage outfalls and the waste of fishing ports.

plumages to aid diving, and thus have to dry out when they come ashore. Cormorants use their feet for propulsion under water, unlike most other diving seabirds which use their wings, though all have webbed feet for surface swimming. Wings used for swimming must be fairly short and paddle-like so they are unsuited to agile flight or gliding.

Penguins have abandoned flight for the most efficient operation in water. Their normal swimming speed when fishing is 5–10 miles per hour (8–16 kilometres per hour) but some species may be able to exceed 40 miles per hour (64 kilometres per hour) when pursued by killer whales. Emperors can dive to 800 feet (240 metres) and stay below for nearly twenty minutes.

Deep diving subjects birds to considerable water pressure and necessitates a stronger structure. The common guillemot and Brunnich's guillemot, which often inhabit the same areas, may not compete with each other because the latter is more robustly built and can dive deeper for its food.

Birds such as gannets and boobies, which plunge in pursuit of prey sighted from the air, have strengthened skulls and foreparts to absorb the shock of impact. They keep their wings spread for course correction until the last moment, folding them back to the shape of a paper dart as they hit the surface. To protect their eyes under water, they close the nictitating membrane – an inner eyelid which all birds possess and flick across to clean the eye or protect it. In diving birds it has a clear central lens so that in effect they wear diving goggles while under water.

Left
The black guillemot or tystie
is found on rocky sea coasts
in the North Atlantic and
North Pacific, nesting in
crevices beneath the
boulders, usually in small,
rather scattered colonies. As
with some other seabirds,
black guillemots will
sometimes work apparently
in concert to round up a
shoal of small fish.

Below
The waved albatross is the
only species to live close to
the equator, where about
12 000 pairs breed in a single
colony at Hood Island in the
Galapagos. This picture
shows the bird's substantial
beak for dealing with fish
and squid.

Above
During courtship display, the male frigate bird inflates a throat sac to enormous size. Frigate birds live on the subtropical and tropical coasts of America, picking fish from the surface without alighting and pursuing other seabirds to force them to disgorge their catches, which are often snatched in midair.

To prevent water being forced into the lungs, the nostrils of gannets and some other seabirds open inside the beak, the external passages being blocked. The gland which removes excess salt from seabirds' bodies and which normally opens through the nostrils is also relocated; the birds open their beaks and shake their heads to disperse the salt.

Beak shapes vary to suit the hunting technique and size of prey. Most are long and pointed for grasping and some have hooked tips as an added aid. The prions have sieve-like ridges inside the beak through which they can expel water while retaining plankton. Most fish eaters swallow their prey whole. Albatrosses which break up large squid, and gulls, skuas and giant petrels which tackle carrion and kill small mammals and other birds, have strong, hook-tipped beaks.

Left
A drake and two duck
common scoter on the shore.
Though they may come
inland into Eurasian and
Alaskan tundra to breed (and
small numbers nest in
Britain), scoters are
essentially marine birds,
diving for shellfish and
wintering in large flocks.

Below
Shags tend to nest in cavities
under boulders and in rock
caves, from which they will
hiss threateningly at an
intruder and display their
brilliant yellow gapes.
Widely distributed around
Europe, shags feed close to
the coast, swimming under
water in pursuit of fish.

When feeding young, some birds carry whole food items back to them in their beaks. Brown pelicans can store captured fish in the distensible pouch of the lower mandible. Puffins have projections inside their beaks to retain what has already been caught while more is added. However, most seabirds carry food to their young by swallowing it and then regurgitating it. While in the stomach it may be partially digested to make it suitable for the chick.

Petrels produce an oily fluid from their food and will squirt it at approaching predators. Probably the primary purpose is simply to lighten themselves for more rapid escape but the fluid is strong-smelling and sticky so that it has a positive defensive function. There are records of fulmars causing the deaths of other birds including falcons by oiling them so badly that their flying ability and plumage insulation is destroyed.

Right
Breeding in Australia and New Zealand, a fairy prion comes ashore at night to incubate its eggs in an underground burrow. These small petrels are nocturnal on land in order to try to avoid the gulls and skuas that hunt them. They feed on plankton which may be more numerous on the surface of the sea at night.

While some seabirds, like penguins and auks, have developed underwater swimming skills with a consequent reduction in flying ability, others have specialized quite differently. Albatrosses and shearwaters have developed long, slim wings on which they can range far and fast using the oceanic wind systems, feeding on or just below the surface of the seas. Like all other petrels, their nostrils are a structure of complicated tubes and it is believed that these can act as pressure-sensing devices so that they can detect

Above
Breeding around the tropical belt, the sooty tern is exceedingly numerous. The feet are only partly webbed and plumage is not waterproof so that, except when nesting, the birds must apparently spend all their time on the wing. Presumably, they are able to sleep in flight, as are some other birds including swifts.

Right
Having probably originated in the Antarctic, fulmars now occur in the North Pacific and North Atlantic and have greatly extended their breeding range in the latter in the last 200 years, becoming common around the coasts of Britain and Iceland and having a scattered distribution in the high Arctic islands.

Left
A young blue-footed booby begs for its parent to regurgitate food. There are six booby species, replacing the gannets in tropical seas. Neither boobies nor gannets have brood patches but incubate their eggs by putting the webs of their feet over them and, in the hottest weather, boobies simply stand beside the egg and shade it from the sun.

Right
The only marine member of its family, the brown pelican breeds in a number of areas on the Pacific coasts of North and South America and around the Caribbean, dispersing more widely along the coasts in winter. Feeding in coastal waters, brown pelicans dive from the air for fish, folding back their wings to a delta shape as they enter the water.

upward air movements caused by waves and exploit them in gliding flight close to the water surface. With their huge narrow wings–the wandering albatross has a span of about 11 feet (3·3 metres)–the larger species cannot fly efficiently by flapping and depend on moving air. They may have to sit out calms and can only nest on island sites where they can launch themselves out to the wind.

Seabirds, such as albatrosses and other petrels, gannets and some auks, may have to range far to sea for their food and can usually only provide for one slow-growing chick. The wandering albatross spends one month preparing to breed, two-and-a-half months incubating and nine months feeding the chick. Thus it takes over a year to produce one offspring and it can only breed in alternate years. Young birds do not usually breed until they are at least six. The low recruitment rate is matched by longevity–some albatrosses may live to be eighty. By contrast, birds like cormorants and terns that feed in coastal waters near the colony can supply food for their young much faster and may rear several chicks at a time each year.

Many seabirds nest colonially. This is partly because of a shortage of suitable sites, which also forces them to go inland. Manx shearwaters breed at 5 000 feet (1 500 metres) in the mountains of Madeira, while great shearwaters and grey shearwaters use the same burrows at different seasons on Tristan da Cunha. For species that nest in the open, the colony also gives some protection from predators, which can be a serious problem. Cliff nesters such as guillemot take their young to sea when they are only half grown and unable to fly properly in order to get them away from the daily attentions of the big gulls and skuas. By contrast, burrow nesters are able to rear their young in safety to the flying stage, finally abandoning them so that hunger forces them to sea. They usually leave at night to avoid hungry eyes.

Colonial species find an added breeding

stimulus in the proximity of other birds and the central sites are always preferred, being occupied by the older birds while those breeding for the first time are forced to the fringes. Despite the apparent attraction of a central location, such birds are still partly territorial. Gannets build their nests just out of beak reach of each other and often strive to steal their neighbours' nesting material. Most gulls hold rather larger territories and boundary disputes are frequent. Each bird has a great psychological advantage on his own ground, and so he will strive to pull the opponent into his territory to give him the beating he deserves. Usually these disputes do not come to serious blows. Anger or frustration is redirected into displacement activities such as tearing at the grass. Many other animals including humans behave similarly and this is an important safety valve for drives that could otherwise harm protagonists.

Gulls will prey on each other's eggs and young as well as those of other birds. The larger species feed extensively in this way, as do great skuas and frigate birds which also steal food from other species, swooping down to harry terns, gannets and boobies in particular until they disgorge. A great skua will even catch a reluctant gannet by the end of one wing and tip it over in the air to make it cooperate. Because of the attentions of these birds, many petrels and others only come ashore at night and remain at sea or in their burrows by day.

Apart from this kleptoparasitism, competition for food is minimized even where several species breed and feed in the same area. The similar cormorant and shag may both be fishing together but the former swims on the bottom so that sixty percent of its catch is flatfish and shrimps, and the latter swims in midwater so that it takes eighty percent sand-eels and herrings. Nor do they compete for nests, as the cormorant prefers the flat tops of islands and stacks but the shag seeks caves and crevices lower down.

Similarly, blue-footed boobies nest on flat ground and feed in inshore shallows, blue-faced boobies feed offshore by day, and the red-footed species nests in bushes and feeds far out at sea, usually at dawn and dusk. Several tern species may nest together and fish by the same method but still not compete because they hunt in different places: little terns close to the beach, common terns rather further out and sandwich terns over deeper offshore water. Where food is superabundant, many species may feed on it at once, but its very quantity means that they are not competing with each other.

At least in the northern hemisphere, seabirds have always been a food resource for man and the opening up of the southern hemisphere produced a similar impact. Most nations now control the level of harvesting. Unfortunately,

Right
The sooty albatross does not nest in colonies as all other species do, but builds a substantial nest cup on a cliff ledge. With little room to move about, the chick remains in the nest until the day it is big enough to fly and can launch itself to space.

Unlike most species, the lava gull has a very dark plumage. Confined to the Galapagos Islands off western South America, it lives on the shores taking a variety of small live prey and carrion. Nesting solitarily, their dark plumage may have some value as a camouflage for the incubating bird.

wherever man went, rats travelled with him; they did and continue to do untold harm to many vast island colonies where ground predators never existed before. Cats and mongooses introduced to control the rats do similar damage.

More modern threats are very diverse and include the use of seabird islands as gunnery or bombing targets, the use of seabird cliffs for rock climbing, and the drowning of large numbers of auks in fixed fishing nets off Greenland. Oil pollution remains a major problem where governments do not take a strong line with oil and tanker companies. Other pollutants such as heavy metals and pesticides are a growing risk.

Guano mining initially caused great damage as many burrowing petrels actually breed in it. It is now managed for a sustained yield but their population must be much lower than it once was.

Gulls have benefited from human waste because far more young birds survive than would naturally, finding easy living at rubbish tips and the like. When they mature and return to the coast to breed their ever-growing numbers affect other seabirds as they take over their nesting spaces and prey on their eggs and young.

Initially whaling and later trawling probably helped such species as fulmar to expand because they could feed on the thousands of tons of waste thrown away at sea. Now factory fishing wastes nothing of the catch—porpoises, pilchards and plankton are all grist to the new fish-meal mills. The effect of this may prove to be more serious than previous human impacts.

Index

Acknowledgments

AFA: Teuvo Suominen 44, 55; D G Allen 96; A B Amerson Jr 182 bottom; Archivio Fotografico Longo 146–7; Ardea: Hans and Judy Beste 76–7, 79, 168, R J C Blewitt endpapers, R Bloomfield 112 top, 126, J B and S Bottomley 54 bottom left, L H Brown 162, Donald D Burgess 83, Elizabeth Burgess 85, 185, R Campbell 124–5 top, K J Carlson 28, 35, 142, Graeme Chapman 74, 82, F Collet 34, Werner Curth 33, John S Dunning 98 top, 101, 107 bottom, M D England 12 bottom, 86, 104, 124–5 bottom, 147, Kenneth W Fink 72, 93 top left, 93 right, 99, 100–1, 120 top, 140 bottom, 178 bottom, John Gooders 26, Clem Haagner front jacket, 186–7, Edgar T Jones 48–9, 51, G Langsbury 176–7, Jens-Peter Laub 178 top, Ake Lindau 153 top, E McNamara 129, P Morris 7, 65, 173, David Pearson 113 bottom, Ralf Richter back jacket flap, B L Sage 45, Peter Steyn front jacket flap, William Stribling 41 bottom, Richard Vaughan 58, John S Wightman 23 bottom, 56 left, 120 bottom right; Atlas Photo: François Merlet 155, 163; Bavaria Verlag: 154 top, N Myers 165 bottom; Arne Blomgren 60 bottom, 61, 62; Bruce Coleman Ltd: H H Barnfather 181, Frank V Blackburn 28–9, 154 bottom, Jane Burton 8 top, Jack Dermid 78, Francisco Erize 116 bottom, Gosta Hakanssen 152 bottom, David Hughes 150, Peter Jackson 130 top, 132, 158, 172, John Markham 20, 76, 91, 107 top, 156, D Middleton 42 bottom, 183, R K Murton 93 bottom left, 96–7, 109, C J Ott 46, John Pearson 110 left, Roger Tory Peterson 153 bottom, 188, Graham Pizzey 108, 117, Goetz D Plage titlespread, S C Porter 72–3, Hans Reinhard back jacket, Leonard Lee Rue III 47, 75, Arne Schmidt 164, Harold Schultz 94–5, Vincent Serventy 134 bottom, James Simon 94, 144 bottom, 156–7, M F Soper 22, 194, Joe van Wormer 52 bottom, 138, 149, 170–1; P R Colston 121; Gerald Cubitt 125; Grahame Dangerfield 113, 141; Evan J Davis 130 bottom, 139 bottom; Hans D Dossenbach 139 top; M P Drazin 116 top; Robert Gillmor 30, 37; Luther Goldman 54 top, 63 top, 136; David A Gowans 81; Jan Grahn 145; Pamela Harrison 24–5; Brian Hawkes 32, 38; George Holton 184; Eric Hosking 11, 59, 130–1, 134 top, 135, 140 top, 148; J E Hunt 70 bottom; Jacana: F Bel and C Vienne 66–7, Michel Brosselin 15, 52 top, 115, 133, 165 top, Jean-Claude Chantelat 21, D Choussy 63 bottom, A R Devez 98 bottom, G Hausle 106, Marc Lelo 120 bottom left, M C Noailles 77, Francis Rave 119 top, Jacques Robert 122, Suinot 39, 40, Philippe Summ 170 bottom, J F and M Terrasse 92, 143, 152 top left, Benoit Tollu 43 top, Jacques Troitignan 137, Varin 6, 118, 175, C Vienne 14, 18, 50 top, 112 bottom, 152 right, 176, Albert Visage 12 top, 23 top, 31; Verna R Johnston 9; Frank W Lane: Ronald Austing 19, 68, 70 top, 71, 89 left, Arthur Christiansen 16, 18–9, 53, 90, 160, Frank W Lane 8 bottom, 13, Georg Quedens 166, R D Robinson 60 top, Dieter Zingel 36; C Laubscher 110 right; H McSweeney 69; W J C Murray 25; NHPA: Stephen Dalton 17, Peter Johnson 103 top, 111, 114, K B Newman 88 top, 119 bottom; Lennart Norstrom 128; Odhams 179; William S Paton 151; Picturepoint 43 bottom; D C H Plowes 89 right; F Quayle 170 top; Betty Risdon 88 bottom; W W Roberts 104–5, 127, 144 top; Paul Schwartz 84; Philippa Scott 41 top, 180; Scott-Swedberg Photos 56–7, 167; Neal G Smith 103; P O Swanberg 54 bottom right, 66; W Tarboton 123; Tierbilder Okapia 87; Thomas W Tivey 161; A D Trounson and M C Clampett 95; Jan Van De Kam 80–1; Ilkka Virkkunen 64; John Warham 42 top, 50 bottom, 182 top; We-Ha Photo, Bern 159, 174; Herbert Zettl 10, 27, 169.

The publishers have made every attempt to contact the owners of the photographs appearing in this book. In the few instances where they have been unsuccessful, they invite the copyright holders to contact them direct.